SpringerBriefs in Applied Sciences and Technology

Series Editor

Andreas Öchsner, Griffith School of Engineering, Griffith University, Southport, QLD, Australia

SpringerBriefs present concise summaries of cutting-edge research and practical applications across a wide spectrum of fields. Featuring compact volumes of 50 to 125 pages, the series covers a range of content from professional to academic.

Typical publications can be:

- A timely report of state-of-the art methods
- An introduction to or a manual for the application of mathematical or computer techniques
- A bridge between new research results, as published in journal articles
- A snapshot of a hot or emerging topic
- An in-depth case study
- A presentation of core concepts that students must understand in order to make independent contributions

SpringerBriefs are characterized by fast, global electronic dissemination, standard publishing contracts, standardized manuscript preparation and formatting guidelines, and expedited production schedules.

On the one hand, **SpringerBriefs in Applied Sciences and Technology** are devoted to the publication of fundamentals and applications within the different classical engineering disciplines as well as in interdisciplinary fields that recently emerged between these areas. On the other hand, as the boundary separating fundamental research and applied technology is more and more dissolving, this series is particularly open to trans-disciplinary topics between fundamental science and engineering.

Indexed by EI-Compendex, SCOPUS and Springerlink.

More information about this series at http://www.springer.com/series/8884

Igor Bolvashenkov · Jörg Kammermann ·
Alexander Rubinraut · Hans-Georg Herzog ·
Ilia Frenkel

Vehicle Electrification

On Water, in Air and Space

 Springer

Igor Bolvashenkov
Institute of Energy Conversion Technology
Technical University of Munich (TUM)
Munich, Bayern, Germany

Alexander Rubinraut
Design Office
Expplanet
Munich, Bayern, Germany

Ilia Frenkel
Center for Reliability and Risk
Management
Shamoon College of Engineering
Beer Sheva, Israel

Jörg Kammermann
Institute of Energy Conversion Technology
Technical University of Munich (TUM)
Munich, Bayern, Germany

Hans-Georg Herzog
Institute of Energy Conversion Technology
Technical University of Munich (TUM)
Munich, Bayern, Germany

ISSN 2191-530X   ISSN 2191-5318   (electronic)
SpringerBriefs in Applied Sciences and Technology
ISBN 978-3-030-81739-8   ISBN 978-3-030-81740-4   (eBook)
https://doi.org/10.1007/978-3-030-81740-4

This Springer imprint is published by the registered company Springer Nature Switzerland AG
The registered company address is: Gewerbestrasse 11, 6330 Cham, Switzerland

Preface

The challenge of vehicular electrification has recently become an extremely important engineering task. This is primarily due to the well-known benefits of electric traction drives. The most significant component in the solution of this problem is the development of highly efficient and fault-tolerant electric propulsion systems that will be the optimal choice for the vehicles, working under specified operating conditions.

The aim of this book is to provide a comprehensive assessment and presentation of various feasible applications of an electric propulsion system, considering their weight, volume, reliability, and fault tolerance. The results of this feasibility analysis can be used today or in the near future for the development of electric propulsion system for the ships, planes, helicopters, and spacecrafts. To solve the above task, we applied new theoretical approaches, including combined random process methods, the Lz-transform technique for multi-state systems, and statistical data processing.

The authors anticipate that the book will be attractive for researchers, practical engineers, and industrial managers in addressing issues related to design and operation of safety-critical traction electric drives. In addition, it will be a helpful textbook for undergraduate and graduate courses in several departments including electrical engineering, industrial engineering, mechanical engineering, and applied mathematics. The book is self-contained and does not require the reader to use other books or papers.

It should be noted that it is impossible to describe all the achievements in the field in a single book. Naturally, some interesting results remained outside of the book's scope. In such cases, the authors provide the readers with the corresponding references.

There are four chapters in this book.

Chapter 1 presents the comprehensive analysis of interrelation between the components reliability features, such as failure rates and repair rates, of a diesel-electric propulsion system of Arctic LNG carrier for the year-round Arctic navigation along the Northern Sea Route, and economic indicators of operational efficiency of icebreaking gas carrier. Arctic navigation imposes specific strict requirements to propulsion systems of the gas tanker, related to the comprehensive reliability and safety. The tasks of implementing year-round navigation along the Northern Sea Route make it even more relevant to ensure reliable and sustainable operations in heavy ice conditions of the eastern sector of the Russian Arctic with the support of atomic icebreakers. The reliability analysis was carried out based on Markov models for multi-state systems.

Chapter 2 is devoted to the feasibility analysis for full-electric aircraft in terms of a conversion design. The advancement of energy storage technologies has given the potential to fully electrify future transport systems. The means of electrifying aircraft are to reduce carbon emission and increase efficiency in air transport. This chapter studies the feasibility of developing an all-electric short to midrange aircraft. The study investigates the possibility of replacing the aircraft's conventional system with an all-electric drive train. Using the current conventional aircraft model, different types of electrical systems are compared to examine their advantages and limitations. The parameters of the turboprop-powered regional airliner De Havilland Canada Dash 8 (DH8D) and the average flight are used to analyze the feasibility of completing a desired range. Additionally, a reliability analysis is conducted for both the conventional and the electrified versions of aircraft, including different electrified topologies.

Chapter 3 is concerned with the investigation of the actual feasibility and prospects of creating an innovative type of electrical helicopter's propulsion system based on battery electric storage, fuel cell electric energy source, ultracapacitors, and a superconducting electric motor for the conventional Airbus helicopter EC135 with two gas turbine engines and a speed reducer. The chapter discusses the advantages and technological problems, which are associated with the thoroughgoing transformation of the conventional propulsion system of helicopter with two turbine engines to the novel electric traction drive topologies. The feasibility evaluation of helicopter's electric propulsion is provided based on the comprehensive analysis, its weight, volume, efficiency, and reliability features.

Chapter 4 focuses on the analysis of the ways to realize the project of electric rockets carrying out flights from Earth's orbit toward planets of the Solar System. The designs, which make it possible to assemble the rocket from individual modules by docking, have been developed. For the jet movement implementation, the electric engine, which belongs to the class of magnetoplasma electric rocket engines, is developed. It has the superconducting exciting winding, the current in which it is directed along the engine axis. That increases the efficiency of the engine significantly. To reduce the flight time, a new technology has been developed which makes it possible to melt ice on the surface of satellites of the planets of the Solar System (Europe, Titan, Oberon, Triton) and using electrolysis to produce hydrogen for tank containers refueling with working substance. A study of the proposed electric propulsion systems was carried out, which allows to optimize the basic parameters of electric engines and onboard electric energy supply systems.

Munich, Germany   Igor Bolvashenkov
Munich, Germany   Jörg Kammermann
Munich, Germany   Alexander Rubinraut
Munich, Germany   Hans-Georg Herzog
Beer Sheva, Israel   Ilia Frenkel

Contents

About the Authors

Igor Bolvashenkov, Ph.D. is Senior Lecturer at the Institute of Energy Conversion Technology of Technical University of Munich (TUM), Munich, Germany. He obtained his M.Sc. (1981) and Ph.D. degrees (1989) in Electrical Engineering from Admiral Makarov State University of Maritime and Inland Shipping, Leningrad, USSR. From 1987 to 1993, he worked as Associate Professor at the Murmansk State Technical University, Russia. Since 2004, he has worked at the Institute of Energy Conversion Technology at the Technical University of Munich (TUM), Munich, Germany.

He specializes in the development and simulation of electric propulsion system for ships, cars, trains, and aircrafts and comprehensive analysis of their efficiency, reliability, and fault tolerance. He has published four books, more than 150 scientific articles, chapters, and patents.

Jörg Kammermann, Dr.-Ing. received his diploma (Dipl.-Ing.) in Electrical Engineering and Information Technology in 2011, as well as his doctoral degree (Dr.-Ing.) in Electrical and Computer Engineering in 2019, from Technical University of Munich (TUM) in Germany. From 2011 to 2016, he was Research Associate, and since 2016, he is Academic Counselor with the Institute of Energy Conversion Technology at TUM.

His research and teaching field includes the system analysis of electric vehicles based on application requirements, multi-phase electric drives, and electric drives for safety–critical applications.

Alexander Rubinraut, Dr.-Ing. Habil is General Manager in the Design Office "Expplanet," (Munich, Germany). Graduated from The Moscow Energetic Institute in 1955, he received a Ph.D. degree in 1967, habilitated in 1989, in Electrical Engineering and Energetics. During 20 years, he was Head of special problem laboratory at the Moscow Research Institute of Electrical Engineering, where electrical motors and generators, operating based on the effect of superconductivity, have been created. He is Author of the book "Cryogenic electrical motors" and two monographs. He has published more than 100 scientific works and invention patents.

Hans-Georg Herzog, Prof. Dr.-Ing. works at the Institute of Energy Conversion Technology, Technical University of Munich (TUM), Munich, Germany. He holds a diploma and doctoral degree (with distinction) from the Technical University of Munich (TUM). After his time as a research associate, he joined Robert Bosch GmbH, Leinfelden-Echterdingen, Germany. Since 2002, he has been Head of the Institute of Energy Conversion Technology at TUM.

His main research interests are energy efficiency of hybrid-electric and full-electric vehicles, electric aircraft, reliability of drive trains and their components, energy and power management, and advanced design methods for electrical machines. He is Senior Member of IEEE and Member of VDI as well as VDE.

Ilia Frenkel, Ph.D. is Chair in the Center for Reliability and Risk Management, Shamoon College of Engineering (SCE), Beer Sheva, Israel.

He obtained his M.Sc. degree in Applied Mathematics from Voronezh State University, Russia, and Ph.D. degree in Operational Research and Computer Science from the Institute of Economy, Ukrainian Academy of Science, Kiev, Ukraine. He has more than 45 years of academic experience and teaching experience at universities and institutions in Russia and Israel. From 1988 till 1991, he worked as Department Chair and Associate Professor at the Applied Mathematics and Computers Department, Volgograd Civil Engineering Institute, Russia. From 2001 till 2018, he served as Senior Lecturer at the Industrial Engineering and Management Department and from 2005 as Chair in the Center for Reliability and Risk Management in Shamoon College of Engineering (SCE), Beer Sheva and Ashdod, Israel.

He specializes in applied statistics and reliability with the application to preventive maintenance. He is Editor and Member of the editorial board of scientific and professional journals. He published five books and more than 100 scientific articles and chapters. He has edited five books and 12 special journal issues and organized several international conferences in Israel and China.

Chapter 1
From the Failure Rate of Components to the Cost-Effectiveness of an Arctic Gas Carrier with an Electric Propulsion System

1.1 Introduction

Presently, due to the large development of gas and oil industry in the Arctic North of Russia, there has been a significant increase in the number of transit traffic along the Northern Sea Route (NSR). This traffic is carried out by large-tonnage tankers by corresponding to the operating conditions ice classes, in the summer–autumn time, supported by atomic icebreakers. In addition, a number of countries are currently considering the NSR as a potentially alternative and competitive option for delivering goods from Asia to Europe and back, bypassing the Suez Canal. This potential is "fueled" by modern data on the melting of Arctic ice in recent years.

Accordingly, the planned year-round navigation and increase the number of shipments along the NSR have increased the need to design and build gas and oil tankers with reinforced Arctic ice classes. In this regard, the propulsion systems of liquefied natural gas (LNG) Arctic tankers are subject to high requirements on operational availability, fault tolerance, and survivability [6, 7, 9, 16, 28], which significantly affect its economic efficiency. The technical data of the tanker are described in detail in the next sections.

1.2 Object of Study

The main task of this study is to determine the dependence of the economic performance of the tanker on the reliability characteristics of the individual components of its propulsion system. To investigate the reliability features of the whole hybrid-electric propulsion system, the modern type of LNG tanker of ice class Arc7 for Arctic navigation "Christophe de Margerie", built in 2017 by Daewoo Shipbuilding & Marine Engineering in South Korea, was chosen. "Christophe de Margerie" is the first of 15 gas tankers built specially for the "Yamal LNG" project. During the winter navigation of 2016–2017, the tanker successfully passed ice tests. The vessel can carry out year-round navigation without icebreaker assistance along the NSR in the western direction and during summer navigation in the eastern direction.

It should be noted that statistics on the operation of this tanker in the eastern sector of the Arctic are currently insufficient. In this regard, statistical data on the operation of approximately similar icebreaker cargo ships for Arctic of the previous generation were used for modeling. Therefore, it can be assumed that the results of this study may be slightly worse than it actually is. With the accumulation of the necessary statistical operational material, the results can be adjusted based on the proposed methodology.

Taking into account that these ships were especially designed for the year-round operation on the NSR, they have special characteristics. The main technical features of LNG tanker "Christophe de Margerie" of the ice class Arc7 are presented in Table 1.1.

Figure 1.1 shows the operating modes of one round trip of LNG tanker "Christophe de Margerie" along the route Sabetta–Shanghai–Sabetta.

During operation cycle depending on external conditions, it is possible to distinguish four basic operating modes of an Arctic LNG tanker. Each of them corresponds to a certain required number and power of the main engines.

They are as follows:

1. Loading and unloading of LNG at the terminal. Sustainability of the loading and unloading process is determined by the reliability of onshore and ship gas liquefying and pumping systems.
2. Navigation of a ship in the ice-free water. Operation in this mode depended on required velocity needs the greater part of the operational time 50%–75% of the nominal generated power.
3. Self-autonomous navigation in the ice without icebreakers support. When navigating in this mode depended on ice conditions, a wide power range of the nominal power can be used.

Table 1.1 Technical data of propulsion system	Specification	Value
	Main diesel engine, MW	11.25×4; 8.45×2
	Traction electric motor, MW	3×15
	Length, width, draft, m	$299 \times 50 \times 11.7$
	Speed in open water, knots	19
	Speed in ice thickness 1.5 m, knots	5.5
	Maximum ice thickness in autonomous navigation, m	2.1
	Cargo capacity, m^3/tons	172,600 / 73,000
	Gross tonnage, tons	130,000
	Number of drives	3
	Ice class	Arc7

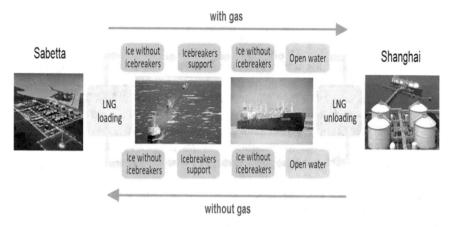

Fig. 1.1 Round trip of LNG carrier on the route Sabetta–Shanghai–Sabetta

4. Navigation of a cargo icebreaker ship in heavy ice supported by nuclear-powered icebreakers (NPIB).

In order to realize sustainable joint operation with atomic icebreakers in this mode, electric propulsion system needs the full available power of diesel generators. In this case, various options for the movement of the ship are possible depending on the severity of the ice conditions. In the first case, it is the construction of a channel in the heavy ice by atomic icebreakers and the independent movement of the tanker behind the icebreakers. In another case, it can be the movement of the tanker in tow with the help of an icebreaker. In the third case, it could be ice fragmentation by icebreakers around a stuck ship and its movement back and forth.

Power demands for various operating conditions for round trip in Asia direction are presented in Fig. 1.2. An especially difficult operating mode is the tanker navigation

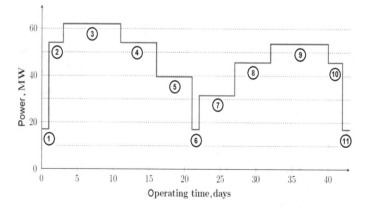

Fig. 1.2 Power demands for various operating conditions for round trip in Asia direction

mode in heavy Arctic ice with the support of two nuclear-powered icebreakers, indicated in Fig. 1.2 by numbers 3 and 9. In these operating modes, the maximum power of the propulsion system and the maximum availability factor are required.

The results of previous study [9] indicate that the first, second, fourth, and fifth operating modes are characterized by the highest possible reliability indicators and do not raise concerns from the point of view of operational safety. This is because relatively small operational power demands are provided with a substantial margin of installed power of multi-power source, as shown in [13, 14].

The minimum durations of these operating modes during the summer–autumn round trip toward Asia and Europe accordingly [4, 5, 22, 29] are shown in Figs. 1.3 and 1.4.

Taking into account the "Novatek"'s plans for year-round use of icebreaking LNG tankers in the eastern sector of the Northern Sea Route, ensuring a sustainable navigating mode with the support of powerful nuclear-powered icebreakers is the most urgent task. In this regard, this study is devoted to a more detailed assessment of this operating mode of the Arctic icebreaking gas carrier, which takes at least 40% of the

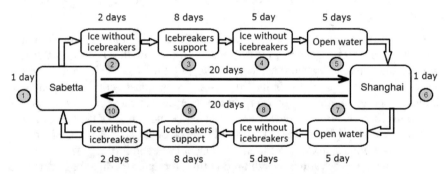

Fig. 1.3 Minimum duration of operational modes during the summer–autumn round trip on the route Sabetta–Shanghai–Sabetta

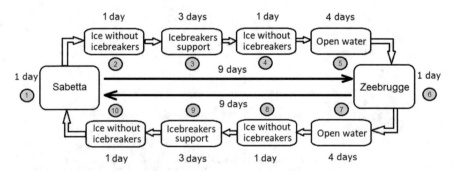

Fig. 1.4 Minimum duration of operational modes during the summer–autumn round trip on the route Sabetta–Zeebrugge–Sabetta

operating time during summer–autumn navigation and requires 100% of the power of propulsion system.

1.3 Methodology of Assessment

For an accurate assessment, the interrelation between reliability indices of propulsion system components, such as failure rate λ and repair rate μ and economic criteria of the icebreaking LNG carrier operation, is proposed to consider and to analyze the entire chain of interactions, from changing the frequency of failures and repairs to the economic effect, depending on these changes. This sequence of interactions is represented by the following expressions.

Estimation of the operational availability of the tanker (A) as function of failure and repair rates (λ, μ) of propulsion system components:

$$A = f(\lambda, \mu), \tag{1.1}$$

Calculation of operational speed of the tanker (V) as function of power of electric propulsion system (N), which is function of operational availability of the tanker (A):

$$V = f(N(A)), \tag{1.2}$$

Determination the economic effect of LNG tanker operation (E) as function of tanker's transport work (the amount of transported LNG), (W), total operational cost (C), which are functions of operational speed of the tanker (V):

$$E = f(W(V), C(V)), \tag{1.3}$$

Here:

  E—economic effect of LNG tanker operation,
  W—tanker's transport work (the amount of transported LNG),
  C—total operational cost,
  V—operational speed of the tanker,
  N—power of electric propulsion system,
  A—operational availability of tanker,
  λ—failure rate of propulsion system components,
  μ—repair rate of propulsion system components.

Firstly, considering the statistical data on the reliability of the propulsion system components, λ and μ the value of operational availability A of the ship is calculated. Appropriate technique is described in detail in the next section. Based on the availability factor, the available shaft power of electric propulsion system of the tanker is determined.

To calculate the average operational velocity of a tanker, it is advisable to use the techniques and formulas for various possible ice navigation conditions given in [1, 4, 5, 15, 20, 23, 25, 27, 28].

To calculate the speed of movement in ice-free water V_0, expression (1.4) is used as follows:

$$V_o = 2,55 \cdot N_{PS}^{1/7,3} \cdot \left(\frac{L}{T}\right)^{1/5} \cdot d^{(-1/6)}, \tag{1.4}$$

where

N_{PS}—power of electric propulsion system in kilowatts,
L—tanker length in meters,
T—tanker draft in meters,
d—tanker hull fullness coefficient.

To calculate the navigational speed in the canal with the support of nuclear-powered icebreakers V_{ICE}, the following expression is used:

$$V_{ICE} = V_0\left[1 - K_{CH}\frac{h_{CH}(V_0 - 2)}{h_{IBS} \cdot V_0}\right] \tag{1.5}$$

where

V_0—speed of movement in ice-free water,
K_{CH}—channel width factor,
h_{CH}—ice thickness in the channel, in meters,
h_{IBS}—thickness of solid smooth ice overcome by the gas carrier in continuous operation at a speed of 2 knots.

The following formula was used to calculate the self-movement velocity of icebreaking tanker in solid ice fields:

$$V_{ICE} = 0.43N_{PS} - 8,8h - 0.01Q \tag{1.6}$$

where

N_{PS}—shaft power of electric propulsion system in in megawatts,
h—thickness of solid ice fields overcome by tanker in meters,
Q—the amount of cargo in thousands of tons.

Figure 1.5 graphically shows the dependence of the velocity of a fully loaded LNG carrier on the shaft power of electric propulsion system for the entire range of ice operating conditions.

As a parameter to evaluate the economic efficiency of the operation of a gas tanker, an indicator of the total operational costs of a chartered ship for one completed circular trip with a full cargo of the gas tanker was adopted, described by the following formula:

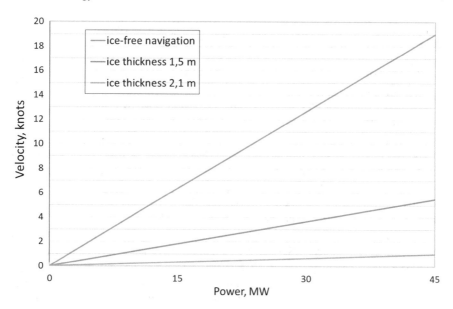

Fig. 1.5 Gas carrier velocity versus power and ice thickness

$$C = C_{FR} + C_{NPIB} + C_{PORT}, \qquad (1.7)$$

where

C_{FR}—LNG carrier freight costs,
C_{NPIB}—the cost of ice channeling support of nuclear-powered icebreakers,
C_{PORT}—port charges costs.

1.4 Model of the LNG Tanker

The structure and the main components of the hybrid-electric propulsion system of the LNG tanker "Christophe de Margerie" are described in detail in [9]. It represents multi-drive, multi-power source, and multi-motor traction system, which can be treated as multi-state system, where components and entire system have an arbitrary finite number of states corresponding to the different performance rates. The performance rate (output nominal power) of the system at any time instant is interpreted as a discrete-state continuous-time stochastic process. In the present chapter, the comparative reliability analyses were carried out by means of Markov models and Lz-transform method. The theoretical basis for the construction and utilization of Markov and Lz-transform methods, the determination of the transition probabilities and reliability evaluation for the hybrid-electric propulsion system that is functioning under various stochastic demands, and its availability and performance are analyzed and presented in [2, 3, 6–8, 10–12, 17–19, 21, 24, 26].

The LNG tanker's power system can be presented as three subsystems: the power supply subsystem, the ship's electric propulsion system, and the subsystem of LNG liquefaction and storage.

The first subsystem, the power supply subsystem (DGSW), includes six diesel generators with a total power of 62,000 kW, which supply electric energy to a two-section main switchboard.

The traction electric drive subsystem (TED) consists of three electric traction drives, including electric converters and three two-section electric traction motors, located in steering gondolas of the Azipod system.

The LNG liquefaction and storage system (LSS) is a critical important consumer, consisting of 12 powerful motor compressors (MCOMP).

In addition to the tanker's power system, two nuclear-powered icebreakers with the same reliable characteristics were considered for tanker's support in heavy ice conditions.

The following is a brief description of the construct the multi-state Markov model of system "tanker icebreakers" defined above. According to [6, 7, 9, 17–19], the Lz-transform method of a discrete-state continuous-time Markov process is used for the following calculations.

We consider that any j-component can have k_j different states, corresponding to different performances g_{ji}, represented by the set $\mathbf{g}_j = \{g_{j1}, ..., g_{jk_j}\}$, $j = \{1, ..., n\}$; $i = \{1, 2, ..., k_j\}$. The performance stochastic processes $G_j(t) \in \mathbf{g}_j$ and the system structure function $G(t) = f(G_1(t), ..., G_n(t))$ that produces the stochastic process corresponding to the output performance of the entire multi-state system, fully define the MSS model.

According to definition [18, 19], Lz-transform of a discrete-state continuous-time Markov process of any j-component $G_j(t)$ is a function defined as follows:

$$L_z\{G_j(t)\} = \sum_{j=1}^{k_j} p_j(t)z^{g_j},$$ (1.8)

where $p_j(t)$ is a probability that the process is in state j at time instant $t \geq 0$ for any given initial states' probability distribution $\mathbf{p}_0$, and z in general case is a complex variable.

Lz-transform from the system structure function $G(t) = f(G_1(t), ..., G_n(t))$ of n independent DSCT Markov processes $G_j(t)$, $j = \{1, ..., n\}$ can be found by applying Ushakov's universal generating operator Ω_f to Lz-transform from $G_j(t)$ processes over all time points $t \geq 0$

$$L_z\{G(t)\} = f(G_1(t), ..., G_n(t)) = \Omega_f(L_z\{G_1(t)\}, ..., L_z\{G_n(t)\}),$$ (1.9)

Figure 1.6 shows the block diagram of the entire LNG tanker's power system, described above.

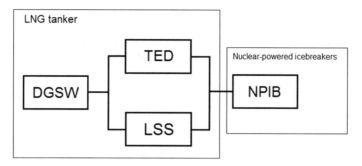

Fig. 1.6 Block diagram of the system "tanker icebreakers"

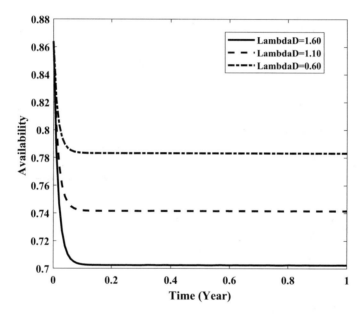

Fig. 1.7 Availability versus failure rates of diesel engine

The structure function of the system "tanker icebreakers" may be presented as follows:

$$G_{TANKER-ICEBREAKES}(t) = \min(G_{DGSW}(t), [G_{TED}(t) + G_{LSS}(t)], G_{NPIB}(t)), \quad (1.10)$$

Applying Ushakov's universal generating operator Ω_f to L_Z-transforms from subsystems of "tanker icebreakers", presented by structure function, the L_Z-transforms to the entire system are as follows:

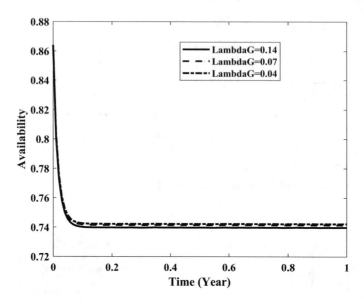

Fig. 1.8 Availability versus failure rates of electric generator

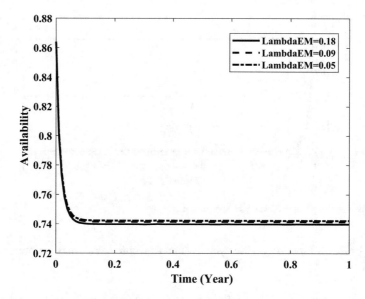

Fig. 1.9 Availability versus failure rates of electric motor

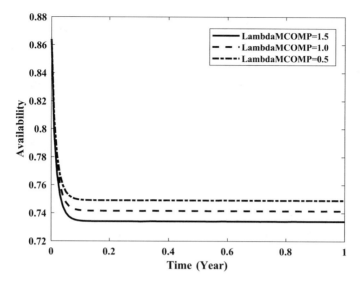

Fig. 1.10 Availability versus failure rates of MCOMP

$$L_z\{G_{TANKER-ICEBREAKES}(t)\} =$$
$$\Omega_{fser}\big(L_z\{G_{DGSW}(t)\}, \Omega_{fpar}(L_z\{G_{DGSW}(t)\}, L_z\{G_{LSS}(t)\}), L_z\{G_{NPIB}(t)\}\big),$$
$$(1.11)$$

The system instantaneous availability $A(t)$ for the constant demand level w is the probability that the multi-state system at instant $t > 0$ is in one of acceptable states:

$$A(t) = \sum_{G(t) \geq w} P_i(t), \qquad (1.12)$$

where $P_i(t)$ is the probability that at instant t the system is in state i.

Corresponding models for subsystems DGSW, TED, and LSS are described in detail in [9].

All subsystem's elements have two states (fully working and fully failed). According to Lz-transform method, in order to calculate the probabilities for each state, we built a state space diagram and the corresponding differential equations.

The failure and repair rates for each system's component considering [6–9] are presented in Table 1.2.

The following graphs 1.7, 1.8, 1.9 and 1.10 show the dependencies of the tanker's operational availability on changes in the reliability characteristics of the components of the propulsion system, which have the greatest influence on the characteristics of the system as a whole. These components are a diesel engine, an electric generator, a main electric motor, and motor–compressors (MCOMP).

Table 1.2 Failure and repair rates of components, year^{-1}

Component	Failure rates	Repair rates
Diesel engine	1,60 / 1.10 / 0,60	54
Generator	0,14 / 0.07 / 0,04	180
Switch	0,10 / 0.05 / 0,03	750
Converter	0,24 / 0.12 / 0,06	659
Transformer	0,28 / 0.14 / 0,07	190
Electric motor	0.18 / 0.09 / 0,05	116
Propeller	0,08 / 0.04 / 0,02	98
Motor–compressor (MCOMP)	1,5 / 1.0 / 0,5	100

Based on the results of availability calculations, the analysis of the remaining parameters of formulas (1.1–1.3) is presented in the next section.

1.5 Results of Calculation

Based on the availability calculation results presented in the previous section and summarized in Table 1.3, a resulting graph of the impact of failure rate values on the operational availability of gas carrier is shown in Fig. 1.11.

For a comparative assessment of the various options, the year-round operation of LNG tankers was considered exclusively eastward, to Korea, China, and Japan. The operational costs of transporting the same amount of LNG during the summer and winter periods of Arctic navigation were compared.

Figures 1.12 and 1.13 show the duration of the round trip and number of possible trips in the summer–autumn and winter–spring periods of navigation, in dependence on the value of reliability characteristics.

Table 1.3 Operational availability of LNG carrier

Component	Failure rates of the motor–compressor (MCOMP)		
	$\lambda_{MCOMP} = 0.5$	$\lambda_{MCOMP} = 1.0$	$\lambda_{MCOMP} = 1.5$
Diesel engine	0.782	0.74	0.702
Generator	0.741	0.74	0.739
Electric motor	0.742	0.74	0.738
Switch	0.741	0.74	0.739
Converter	0.743	0.74	0.737
Transformer	0.744	0.74	0.736
Propeller	0.741	0.74	0.739
MCOMP	0.748	0.74	0.734

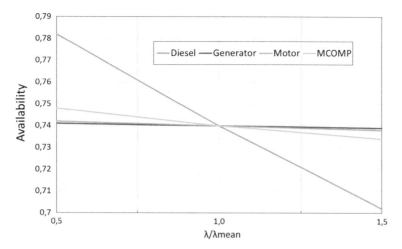

Fig. 1.11 Impact of failure rates of components on the operational availability

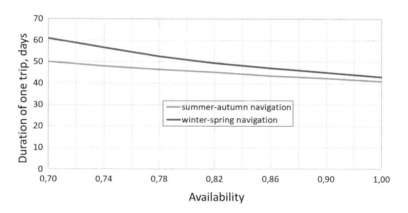

Fig. 1.12 Impact of operational availability on duration of one round trip

The calculations of economic performance were performed using the following costs:

- The tanker freight cost $C_{FR} = 140,000$ USD/day.
- The one icebreaker support cost C_{AIB} were determined on the basis of the Icebreaker Assistance Calculator developed by the NSR Administration.
- The port charges $C_{PORT} = 150,000$ USD for Sabetta and $C_{PORT} = 130,000$ USD for Shanghai.

Table 1.4 presents the results of calculations of the operational costs during one full trip of the Arctic gas tanker eastward in the summer–autumn and winter–spring periods of navigation.

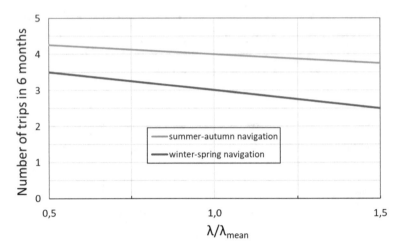

Fig. 1.13 Impact of failure rates of components on the round trips number

Table 1.4 Operational cost for one round trip eastward, USD·10^6

Route of navigation	Season	C_{FRT}	C_{NPIB}	C_{PORT}	Total
Eastward, failure rate 0.5	Summer	6.44	0.55	0.28	7.27
	Winter	7.14	1.40	0.28	8.82
Eastward, failure rate 1.0	Summer	6.72	0.55	0.28	7.55
	Winter	7.98	1.40	0.28	9.66
Eastward, failure rate 1.5	Summer	7.14	0.55	0.28	7.97
	Winter	8.82	1.40	0.28	10.50
$(Cost_{1.5}/Cost_{0.5})$	Summer	1.11	1.00	1.00	1.10
	Winter	1.24	1.00	1.00	1.19
$(Cost_{1.5} - Cost_{0.5})$	Summer	–	–	–	0.70
	Winter	–	–	-	1.68

A graphical representation of the change in the value of operational costs depending on the failure rates of the main components of the propulsion system of LNG tanker is shown in Fig. 1.14.

As can be seen from Fig. 1.14, the reliability characteristics of the components have a significant impact on the economic efficiency of the operation of the ship. At the same time, the influence of various components differs significantly from each other, which should be taken into account when designing ship propulsion systems for Arctic navigation.

The projected annual gas production by "Novatek" from the Yamal natural gas field is 20 million tons of gas per year. To transport this volume of liquefied gas, 15 gas carriers with Arc7 ice class were built. There are two alternative routes to transport LNG from the Yamal Peninsula: westward, to the European market or eastward, to the

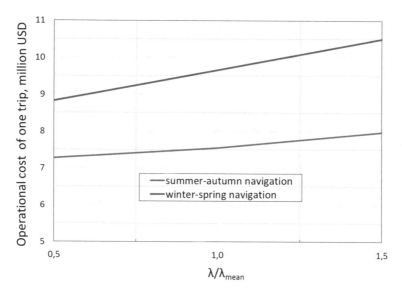

Fig. 1.14 Impact of failure rates of components on the operational costs

Asian market: to China, Japan, and Korea. By 2022, "Novatek" plans to deliver LNG to Asia only via the Northern Sea Route, while building a transshipment complex in Kamchatka.

In the future, to reduce shipping costs, LNG will be reloaded from icebreaker-class tankers to conventional tankers in order to reduce the overall operating costs of transportation to Asian markets.

The results of the analysis of the potential for transporting 20 million tons of LNG during one year are presented in Fig. 1.15. The horizontal axis shows the operating time of the LNG carrier to the east direction $t_{eastwards}$.

Accordingly, the operating time of a gas carrier for the European market $t_{westwards}$ is as follows:

$$t_{westwards} = 12 - t_{eastwards}$$

From Fig. 1.15, it can be seen that achieving the specified annual volume of LNG transportation with 15 LNG carriers is possible only with year-round operation of vessels in the direction of Europe. With the planned future operation of gas carriers exclusively to the east direction, the number of Arctic gas carriers should be significantly increased.

Table 1.5 summarized the data of the calculations of the annual operating profit of the gas carrier when operating in the direction of Europe for various market prices of LNG.

One cubic meter of LNG becomes 600 cubic meters of natural gas when vaporized back to its gaseous form. Thus, the total volume of gas transported by the "Christophe de Margerie" type gas tanker in one trip is approximately 100 million cubic meters

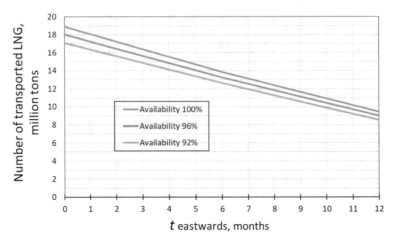

Fig. 1.15 Number of LNG transported by 15 gas carriers in one year

Table 1.5 Annual operational profit for the westward trips, mil. USD

LNG price, USD/1000 m^3	Availability (%)	Trips number	Income, mil.USD	Costs, mil.USD	Profit, mil.USD
250	100	18	450	383.4	66.6
	92	17	425	362.1	62.9
350	100	18	630	383.4	246.6
	92	17	595	362.1	232.9
450	100	18	810	383.4	426.6
	92	17	765	362.1	402.9

of gas. Taking into account the data of "Novatek," the cost of gas from the "Yamal LNG" field when supplied to the European market is $213 per thousand cubic meters. At the same time, when LNG is supplied to Asia, the cost is $273.35 per thousand cubic meters. Table 1.6 summarized the data of the calculations of annual operating

Table 1.6 Annual operational profit for the eastward trips, millions USD

LNG price, USD/1000m^3	Availability (%)	Trips number	Income, in MM$	Costs, mil.USD	Profit, mil.USD
250	100	9	225	245.7	−20.7
	92	8	200	218.4	−18.4
350	100	9	315	245.7	69.3
	92	8	280	218.4	61.6
450	100	9	405	245.7	159.3
	92	8	360	218.4	141.6

profit of the gas carrier when operating in the direction of Asia for various market prices of LNG.

In Figs. 1.16 and 1.17, the comparative results of the annual operating profit for the operation of LNG carriers westward and eastwards, respectively, are presented graphically.

Comparative analysis of the diagrams indicates a significantly higher profitability of gas carrier trips to the Europe. At the same time, the transportation of LNG in the Asian direction becomes profitable when the market price of gas exceeds 300 USD per 1000 m^3. The impact of availability on annual profit increases with an increase in the market price of LNG.

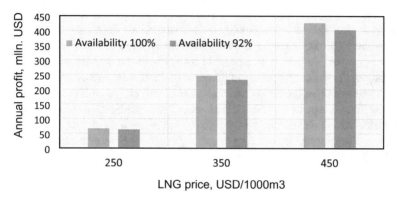

Fig. 1.16 Annual operating profit for the operation of LNG carriers westward

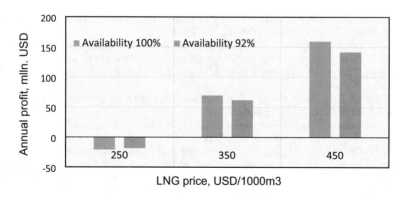

Fig. 1.17 Annual operating profit for the operation of LNG carriers eastward

1.6 Conclusion

The proposed technique allows to evaluate the interdependencies between the simple reliability characteristics of the basic components and units of the electric propulsion system of the Arctic gas carriers on the operational economic efficiency of their use on the NSR, as well as to propose the recommendations for its optimization. The preliminary estimation of the operational availability and economic indicators of the application of hybrid diesel-electric power system of the Arctic LNG gas carrier "Christophe de Margerie" for year-round navigation westward and eastward of the NSR is carried out.

Based on the results of the study, it is possible to predict the profitability of individual trips regarding the current situation on the global LNG market and the appropriate direction of navigation, as well as to forecast the required number of gas carriers and icebreakers for the required volume of cargo transportation.

In the further studies, it is advisable to estimate the impact on the economic efficiency of the gas carrier, the aging process of the components, the deterioration of the ice overcoming of the tanker and icebreakers, and various maintenance strategies of propulsion system. In addition, it is necessary to consider the optimization of east–west traffic routes depending on the season, ice conditions and world LNG prices market, and the possibility of using reloading terminals to increase the coefficient of use of LNG tankers in ice conditions.

References

1. R. Abbassi, F. Khan, N. Khakzad et al., Risk analysis of offshore transportation accident in arctic waters. Int. J. Marit Eng. **159**(A3), 213–224 (2017)
2. M. Altosole, U. Campora, S. Savio, Improvements of the ship energy efficiency by a steam powered turbogenerator in LNG propulsion applications, in *Proc. of International Symposium on Power Electronics, Electrical Drives, Automation and Motion (SPEEDAM)*, (Amalfi, Italy, 2018), p. 7
3. A. Barabadi, T. Markeset, Reliability and maintainability performance under Arctic conditions. Int. J. Syst. Assur. Eng. Manag **2**(3), 205–217
4. I. Bolvashenkov, *Impact of operational conditions of icebreaker ships on the optimal type of propulsion system* (Nova Science, New-York, 2018)
5. I. Bolvashenkov, H.G. Herzog, Use of stochastic models for operational efficiency analysis of multi power source traction drives, in *Proc. of the Second IEEE International Symposium on Stochastic Models in Reliability Engineering, Life Science and Operations Management, (SMRLO)*, ed. By I. Frenkel, A. Lisnianski (Beer Sheva, Israel, 2016), pp. 124–130
6. I. Bolvashenkov, H.G. Herzog, I. Frenkel et al., *Safety-Critical Electrical Drives: Topologies, Reliability, Performance* (Springer, Switzerland, 2018)
7. I. Bolvashenkov, H.G. Herzog, F. Ismagilov et al., *Fault-Tolerant Traction Electric Drives: Reliability, Topologies and Components Design* (Springer, Singapore, 2020)
8. I. Bolvashenko, J. Kammermann, H.G. Herzog, Methodology for determining the transition probabilities for multi-state system markov models of fault tolerant electric vehicles. In *Proc. of Int. Asian Conference on Energy, Power and Transportation Electrification (ACEPT'16)*, Singapore, 2016, p. 6

9. I. Bolvashenkov, J. Kammermann, I. Frenkel et al., Operational availability and performance analysis of the multi-drive multi-motor electric propulsion system of an icebreaker gas tanker for arctic. in *Proc. of IEEE 14th International Conference on Ecological Vehicles and Renewable Energies (EVER'19)*. Monaco, 8–10 Mai 2019, pp. 1–6

10. I. Bolvashenkov, J. Kammermann, I Frenkel et al., Fault tolerance assessment of multi-motor electrical drives with multi-phase traction motors based on *Lz*–transform, in *Proc. of IEEE 14th International Conference on Ecological Vehicles and Renewable Energies (EVER'19)*, 8–10 Mai 2019, Monaco, pp. 1–6.

11. G.G. Dimopoulos, C.A. Frangopoulos, Optimization of propulsion systems for modern LNG carriers considering multiple technology and design alternatives, in *Proc. of 10th International Marine Design Conference (IMDC)*, Trondheim, Norway, 26–29 May 2009, p. 20

12. I.A. Fernándeza, M.R. Gómez, J.R. Gómez et al., Review of propulsion systems on LNG carriers. Renew. Sustain. Energy Rev. **67**, 1395–1411 (2017)

13. I. Frenkel, I. Bolvashenkov, H.G. Herzog et al., Performance availability assessment of combined multi power source traction drive considering real operational conditions. Transp. Telecommun. **17**(3), 179–191 (2016)

14. I. Frenkel, I. Bolvashenkov, H.G. Herzog et al., Operational Sustainability Assessment of Multipower Source Traction Drive, in *Mathematics Applied to Engineering*. ed. by M. Ram, P. Davim (Elsevier, Academic Press, UK, 2017), pp. 191–203

15. F. Goerlandt, J. Montewka, W. Zhang et al., An analysis of ship escort and convoy operations in ice conditions. Saf. Sci. **95**, 198–209 (2017)

16. T. Huan, F. Hongjun, L. Wei et al., *Options and Evaluations on Propulsion Systems of LNG Carriers, Propulsion Systems*. (IntechOpen, 2019), pp.1–20

17. A. Lisnianski, Lz-Transform for a Discrete-State Continuous-Time Markov Process and its Application to Multi-State System Reliability, in *Recent Advances in System Reliability*. ed. by A. Lisnianski, I. Frenkel (Springer-Verlag, London, 2012), pp. 79–95

18. A. Lisnianski, I. Frenkel, Y. Ding, *Multi-State System Reliability Analysis and Optimization for Engineers and Industrial Managers* (Springer, London, 2010)

19. A. Lisnianski, I. Frenkel, L. Khvatskin, *Modern Dynamic Reliability Analysis for Multi-state Systems. Springer Series in Reliability Engineering* (Springer, Cham, 2021)

20. N. Marchenko, *Russian Arctic Seas. Navigational Conditions and Accidents* (Springer, Berlin, Heidelberg, 2012)

21. R. Pfaff, K. Melcher, J. Franzen, Rare event simulation to optimize maintenance intervals of safety critical redundant subsystems, in *Proc. of European Conference of the Prognostics and Health Management Society (PHME'18)*, Utrecht, Netherlands, 3–6 July 2018, p. 6

22. C.K. Pil, M. Rausand, J. Vatn, Reliability assessment of reliquefaction systems on LNG carriers. Reliab. Eng. Syst. Saf. **93**, 1345–1353 (2008)

23. V.V. Plotnikov, V.I. Pustoshnova, Variability and conjugacy of ice conditions in the system of East Arctic Seas (the Laptev, East Siberian, and Chukchi Seas). Russ. Meteorol. Hydrol. **37**(7), 468–476 (2012)

24. M.R. Pourhassan, S. Raissi, A. Hafezolkotob, Effects of faulty estimate in component reliability on system designing: a simulation approach. J. Ind. Syst. Eng. **12**(2), 174–185 (2019)

25. K. Riska, Ship – ice interaction in ship design: theory and practice, in *Encyclopedia of Life Support Systems. Cold Regions Science and Marine Technology, (UNESCO-EOLSS)*, p. 30

26. V. Romanovskiy, M. Sjubaev, I. Bolvashenkov, Selection basic data of electrical machines for electrical propulsion systems, in *Bulletin of the State University of Maritime and River Fleet*, 6(34), (Sankt Petersburg, Russia, 2015), pp. 172–178

27. O.V. Sormunen, R. Berglund, M. Lensu et al., Comparison of Vessel Theoretical Ice Speeds Against AIS Data in the Baltic Sea, in *Marine Design XIII*. (Taylor & Francis Group, London, 2018), pp. 841–849

28. L.G. Tsoy, Study of ice characteristics and justification of rational parameters of the ships for ice navigation, (Nestor-Historia, Sankt-Petersburg, Russia, 2017)

29. C. Vianello, G. Maschio, Risk analysis of LNG terminal: case study. Chem. Eng. Trans. **36**, 277–282 (2014)

Chapter 2
Technological Feasibility of a Full-Electric Aircraft Considering Weight, Volume, and Reliability Restrictions

2.1 Introduction

With the ongoing goal to reduce the environmental impact of transport vehicles, there have been developments in hybrid-electric and all-electric vehicles. In the success of commercializing hybrid-electric and all-electric cars and continuing improvements in energy storage technologies, there is a potential to improve on the electrification of commercial rail vehicles, watercraft, and aircraft. This fact and the societal push to reduce carbon emission have increased the interest to electrify transport vehicles. With the trend of increasing air travels and 85% of the flights are short haul flights [16], the development of electrifying electric aircraft to reduce emission is crucial.

In addition to reducing emissions, electrifying conventional systems generally allows for higher reliability and efficiency, lower maintenance cost, and a reduction in noise, weight, and fuel consumption [1, 9]. Producing an all-electric aircraft is proven possible with for instance the airbus's E-Fan 1.0 [6]. However, the electrification of some transport vehicles has better advantages than others based on the application. In comparison with other vehicles, the aircraft's weight is one of the key factors in its operation. Therefore, it is essential to determine whether the priorities of fully electrifying aircraft over other vehicles are beneficial. The evaluation from [15] shows that aircraft are beneficial for long-range and oceanic flight, but in terms of short-range flights, they are only advantageous for flights to areas not accessible by high-speed rail vehicles and when the cost of infrastructure is lower than of high-speed rail vehicles. The studies on electrifying aircraft have raised with growing demand in air travel, for instance [2, 4, 5, 7, 13, 14]. The electrification of aircraft can increase efficiency and offer new applications such as vertical takeoff and landing which would reduce the cost of infrastructure. Therefore, the development of a hybrid-electric or an all-electric aircraft is highly favorable. However, the challenge arises with the range and current state of energy storage technology. Since mass is an important factor in aircraft design, the energy density of the storage technology is a crucial factor; other components in the drive train are also evaluated since they are essential to the electric propulsion system [3].

© The Author(s), under exclusive license to Springer Nature Switzerland AG 2022
I. Bolvashenkov et al., *Vehicle Electrification*, SpringerBriefs in Applied Sciences and Technology, https://doi.org/10.1007/978-3-030-81740-4_2

This chapter incorporates the methodologies discussed in literature to analyze the conversion of a small-range turboprop aircraft to an all-electric aircraft (AEA). The conventional propulsion system is replaced with an electric propulsion system with the parameters of the electrical components taken from literature; the rest of the system and aircraft parameters are unchanged. The conversion of the aircraft will evaluate the feasibility of integrating the electric propulsion system given the weight restrictions. Battery and fuel cell storage technologies are analyzed and compared to the conventional aircraft system by their range performance and by reliability. The preliminary investigations of this task were carried out in [10] and are developed further in detail within this chapter.

2.2 Analysis of Weight and Volume

In this chapter, the parameters of the turboprop-powered regional airliner De Havilland Canada Dash 8 (DH8D) are used for the conversion of the conventional propulsion system to an electric propulsion system. Kerosene fuel is fed to the engine which drives the propeller for the conventional propulsion system. The drive train for the all-electric aircraft replaces the conventional propulsion system with a supply system for electric energy, converter, and motor. The electric propulsion system will be solely inspected based on its specific energy density and power density; all other components and factors are omitted. Furthermore, the flight profile of the aircraft is held under normal conditions. The electrical conversion of the propulsion system for the De Havilland Canada Dash 8 evaluates the feasibility of fully electrifying a short-range 90 passenger aircraft. Given that the energy density of kerosene is much higher compared to the energy density of electrical storage systems, the conversion of the aircraft is optimized based on the fuel weight while other necessary weights, such as fuel cell and converter weight, are transferred to the payload; the size of the payload is adjusted such that the fuel storage is at its maximum capacity and the total weight does not exceed the maximum takeoff weight (MTOW).

2.2.1 Aircraft Parameters

Hence, the MTOW and maximum landing weight (MLW) are used, respectively, as the upper and lower weight limits for the calculation. In the case of an AEA, with an exclusive use of a battery storage system, the MLW is taken as MTOW since the fuel weight does not change throughout the flight. Table 2.1 provides the specification of the DH8D [10].

Table 2.1 System parameters and technical data of the DH8D [10]

Parameter	Value
Maximum takeoff weight (MTOW)	30,481 kg
Maximum landing weight (MLW)	28,010 kg
Operating empty weight (OEW)	17,819 kg
Engine weight (PW150A)	717 kg
Maximum payload	8489 kg
Maximum fuel capacity	5220 kg
Maximum engine power	3781 kW
L/D ratio	15.5

2.2.2 Aircraft Conversion Analysis

Without cost considerations, the reasons to build all-electric aircrafts must have benefits that outweigh the other types of all-electric vehicles. Some practical needs would be for long-range flights or the capability to reach areas not accessible with land vehicles. The range equations given in [8] are applicable for the electric propulsion systems studied. More detailed information can be found in [10].

For the analysis of the hybrid battery and fuel cell system, the parameters of Table 2.2 are considered. The fuel capacity (5220 kg) is partitioned between battery and hydrogen and tank. In order to optimize on the maximum payload weight and maintain the maximum landing weight (MLW) limit, the initial fuel system will have 2500 kg of the hydrogen. All of the hydrogen will be burned so that the aircraft's weight will be within the range of the MLW. In this method, the hydrogen fuel is used before the battery so that the aircraft weight is lighter and will provide more range for the battery system; with the hydrogen initially occupying 2500 kg of the fuel weight, the initial battery weight is 2720 kg. The system is evaluated based on the range of the battery. After burning the hydrogen, the range equation will compare the range under the condition that the remaining fuel weight is either hydrogen or

Table 2.2 Weight of fuel cell and battery propulsion components

Component	Weight	
	Fuel cell	Battery
Converter	374.0 kg	
Fuel cell	3113.4 kg	–
Hydrogen and tank	5220.0 kg	–
Battery	–	5220.0 kg
Gearbox	1744.9 kg	
Motor	1140.0 kg	
New operating empty weight	19,270.1 kg	
New maximum payload	2503.5 kg	3145.9 kg

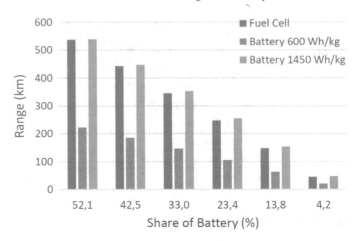

Fig. 2.1 Battery–fuel cell hybrid range comparison with battery energy density of 600 Wh/kg and of 1450 Wh/kg

battery weight. If the fuel cell range exceeds the battery range, an additional 500 kg of hydrogen is used of the shared fuel weight while the battery capacity is reduced by 500 kg. The 500 kg of hydrogen is then burned, reducing the aircraft's mass, and the battery and fuel cell's ranges are compared again. This process repeats, with the share of hydrogen increasing, until the battery range exceeds the hydrogen range for the hybrid system.

The results of the hybrid system are shown in Fig. 2.1. The figure shows that having a hybrid system will reduce the overall range of the aircraft with the given fuel capacity. With a system consisting of 95.8% hydrogen, with all the hydrogen initially burned, the chart shows that using the last 4.2% of the fuel weight for hydrogen would yield more distance than for battery. If the fuel limit is increased and the aircraft is still operational, the use of battery can be advantageous for the hybrid system. However, with the current fuel capacity maintained for an operational aircraft, the battery's energy density must increase to have a beneficial hybrid system. Figure 2.1 shows that the range of the battery starts to exceed the range of hydrogen when the energy density of the battery is about 1450 Wh/kg.

2.3   Reliability and Fault Tolerance Assessment of the DH8D

The reliability block diagram of the conventional aircraft with two, four, and six fuel pumps, is presented in Figs. 2.2, 2.3, and 2.4, respectively.

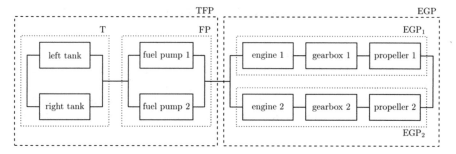

Fig. 2.2 Reliability block diagram of the conventional aircraft with two fuel pumps

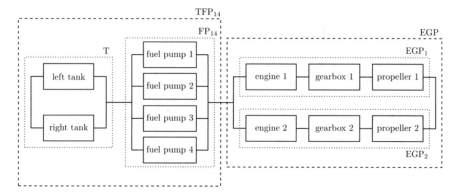

Fig. 2.3 Reliability block diagram of the conventional aircraft with four fuel pumps

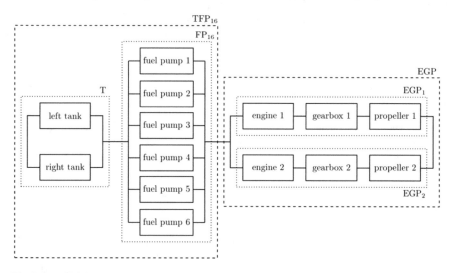

Fig. 2.4 Reliability block diagram of the conventional aircraft with six fuel pumps

2.3.1 Description of the System's Elements

All system's elements, left tank (LT), right tank (RT), left fuel pump (LFP), right fuel pump (RFP), fuel pumps (FP$_1$-FP$_6$), engines (E$_1$, E$_2$), gearboxes (G$_1$, G$_2$), and propellers (P$_1$, P$_2$), have two states: fully working and fully failed. According to Lz-transform method [11, 12], in order to calculate the probabilities for each state, we built for all system's elements the state space diagram (Fig. 2.5) and the following system of differential equations:

$$\begin{cases} \dfrac{dp_1^{(i)}(t)}{dt} = -\lambda_i p_1^{(i)}(t) + \mu_i p_2^{(i)}(t), \\ \dfrac{dp_2^{(i)}(t)}{dt} = \lambda_i p_1^{(i)}(t) - \mu_i p_2^{(i)}(t). \end{cases} \tag{2.1}$$

$i = $ LT, RT, LFP, RFP, E$_1$, E$_2$, G$_1$, G$_2$, P$_1$, P$_2$.

Initial conditions are as follows: $p_1^{(i)}(0) = 1$; $p_2^{(i)}(0) = 0$.

We used MATLAB for the numerical solution of these systems of differential equations to obtain probabilities $p_1^{(i)}(t)$, $p_2^{(i)}(t)$, ($i = $ LT, RT, LFP, RFP, E$_1$, E$_2$, G$_1$, G$_2$, P$_1$, P$_2$).

Therefore, for elements of such system, the output performance stochastic processes and corresponding Lz-transforms can be defined as follows:

For $i = $ LT, RT, LFP, RFP, E$_1$, E$_2$, G$_1$, G$_2$, P$_1$, P$_2$

$$\begin{cases} \mathbf{g}^{(i)} = \left\{ g_1^{(i)}, g_2^{(i)} \right\} = \{50, 0\} \\ \mathbf{p}^{(i)} = \left\{ p_1^{(i)}(t), p_2^{(i)}(t) \right\}. \end{cases}$$

$$L_z\{G^{(i)}(t)\} = p_1^{(i)}(t) z^{g_1^{(i)}} + p_2^{(i)}(t) z^{g_2^{(i)}} = p_1^{(i)}(t) z^{50} + p_2^{(i)}(t) z^0 \tag{2.2}$$

For $i = $ FP$_1$-FP$_6$

Fig. 2.5 State space diagram

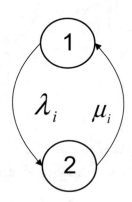

$$\begin{cases} \mathbf{g}^{(i)} = \left\{g_1^{(i)}, g_2^{(i)}\right\} = \{25, 0\} \\ \mathbf{p}^{(i)} = \left\{p_1^{(i)}(t), p_2^{(i)}(t)\right\}. \end{cases}$$

$$L_z\left\{G^{(i)}(t)\right\} = p_1^{(i)}(t)z^{g_1^{(i)}} + p_2^{(i)}(t)z^{g_2^{(i)}} = p_1^{(i)}(t)z^{25} + p_2^{(i)}(t)z^0 \qquad (2.3)$$

For simplification, we calculate output performance in percent.

2.3.2 Lz-Transform Calculation of the Multi-state Models for the Conventional Aircraft

2.3.2.1 Conventional Aircraft with Two Fuel Pumps

According to Fig. 2.2 and using the recursive derivation approach [11, 12], we will present the whole system using the Lz-transform as follows:

$$\begin{aligned} L_z\left\{G^T(t)\right\} &= \Omega_{f_{par}}\left(L_z\left\{g^{LT}(t)\right\}, L_z\left\{g^{RT}(t)\right\}\right) \\ L_z\left\{G^{FP}(t)\right\} &= \Omega_{f_{par}}\left(L_z\left\{g^{LFP}(t)\right\}, L_z\left\{g^{RFP}(t)\right\}\right) \\ L_z\left\{G^{TFP}(t)\right\} &= \Omega_{f_{ser}}\left(L_z\left\{G^T(t)\right\}, L_z\left\{G^{FP}(t)\right\}\right), \\ L_z\left\{G^{EGP_1}(t)\right\} &= \Omega_{f_{ser}}\left(L_z\left\{g^{E_1}(t)\right\}, L_z\left\{g^{G_1}(t)\right\}, L_z\left\{g^{P_1}(t)\right\}\right), \\ L_z\left\{G^{EGP_2}(t)\right\} &= \Omega_{f_{ser}}\left(L_z\left\{g^{E_2}(t)\right\}, L_z\left\{g^{G_2}(t)\right\}, L_z\left\{g^{P_2}(t)\right\}\right), \\ L_z\left\{G^{EGP}(t)\right\} &= \Omega_{f_{par}}\left(L_z\left\{G^{EGP_1}(t)\right\}, L_z\left\{G^{EGP_2}(t)\right\}\right) \\ L_z\left\{G^{Conv2FP}(t)\right\} &= \Omega_{f_{ser}}\left(L_z\left\{G^{TFP}(t)\right\}, L_z\left\{G^{EGP}(t)\right\}\right). \end{aligned} \qquad (2.4)$$

Using the composition operators $\Omega_{f_{ser}}$, $\Omega_{f_{par}}$ for subsystems, we obtain the following Lz-transforms:

- Lz-transforms for tanks (T) subsystem:

$$\begin{aligned} L_z\left\{G^T(t)\right\} &= \Omega_{f_{par}}\left(L_z\left\{g^{LT}(t)\right\}, L_z\left\{g^{RT}(t)\right\}\right) \\ &= P_1^T(t)z^{100} + P_2^T(t)z^{50} + P_3^T(t)z^0. \end{aligned} \qquad (2.5)$$

where

$$\begin{aligned} P_1^T(t) &= p_1^{LT}(t) \cdot p_1^{RT}(t) \\ P_2^T(t) &= p_1^{LT}(t) \cdot p_2^{RT}(t) + p_2^{LT}(t) \cdot p_1^{RT}(t) \\ P_3^T(t) &= p_2^{LT}(t) \cdot p_2^{RT}(t) \end{aligned}$$

- *Lz*-transforms for fuel pumps (*FP*)subsystem:

$$L_z\{G^{FP}(t)\} = \Omega_{f_{par}}\left(L_z\{g^{LFP}(t)\}, L_z\{g^{RFP}(t)\}\right)$$
$$= P_1^{FP}(t)z^{100} + P_2^{FP}(t)z^{50} + P_3^{FP}(t)z^0. \tag{2.6}$$

where

$$P_1^{FP}(t) = p_1^{LFP}(t) \cdot p_1^{RFP}(t)$$
$$P_2^{FP}(t) = p_1^{LFP}(t) \cdot p_2^{RFP}(t) + p_2^{LFP}(t) \cdot p_1^{RFP}(t)$$
$$P_3^{FP}(t) = p_2^{LFP}(t) \cdot p_2^{RFP}(t)$$

- *Lz*-transforms for tanks–pumps (*TFP*) subsystem:

$$L_z\{G^{TFP}(t)\} = \Omega_{f_{par}}\left(L_z\{G^T(t)\}, L_z\{G^{FP}(t)\}\right)$$
$$= P_1^{TFP}(t)z^{100} + P_2^{TFP}(t)z^{50} + P_3^{TFP}(t)z^0. \tag{2.7}$$

where

$$P_1^{TFP}(t) = P_1^T(t) \cdot P_1^{FP}(t)$$
$$P_2^{TFP}(t) = P_1^T(t) \cdot P_2^{FP}(t) + P_2^T(t) \cdot P_1^{FP}(t) + P_2^T(t) \cdot P_2^{FP}(t)$$
$$P_3^{TFP}(t) = P_1^T(t) \cdot P_3^{FP}(t) + P_2^T(t) \cdot P_3^{FP}(t) + P_3^T(t) \cdot P_1^{FP}(t)$$
$$+ P_3^T(t) \cdot P_2^{FP}(t) + P_3^T(t) \cdot P_3^{FP}(t)$$

- *Lz*-transforms for engines–gearboxes subsystem:

$$L_z\{G^{EG_i}(t)\} = \Omega_{f_{ser}}\left(L_z\{g^{E_i}(t)\}, L_z\{g^{G_i}(t)\}\right)$$
$$= P_1^{EG_i}(t)z^{50} + P_2^{EG_i}(t)z^0, \, i = 1, 2 \tag{2.8}$$

where

$$P_1^{EG_i}(t) = p_1^{E_i}(t) \cdot p_1^{G_i}(t)$$
$$P_2^{EG_i}(t) = p_1^{E_i}(t) \cdot p_2^{G_i}(t) + p_2^{E_i}(t) \cdot p_1^{G_i}(t) + p_2^{E_i}(t) \cdot p_2^{G_i}(t)$$

- *Lz*-transforms for engines–gearboxes–propellers subsystem:

$$L_z\{G^{EGP_i}(t)\} = \Omega_{f_{ser}}\left(L_z\{G^{EG_i}(t)\}, L_z\{g^{P_i}(t)\}\right)$$
$$= P_1^{EGP_i}(t)z^{50} + P_2^{EGP_i}(t)z^0, i = 1, 2 \tag{2.9}$$

where

$$P_1^{EGP_i}(t) = P_1^{EG_i}(t) \cdot p_1^{P_i}(t)$$

$$P_2^{EGP_i}(t) = P_1^{EG_i}(t) \cdot p_2^{P_i}(t) + P_2^{EG_i}(t) \cdot p_1^{P_i}(t) + P_2^{EG_i}(t) \cdot p_2^{P_i}(t)$$

- Lz-transforms for *EGP* subsystem:

$$L_z\{G^{EGP}(t)\} = \Omega_{f_{par}}\left(L_z\{G^{EGP_1}(t)\}, L_z\{G^{EGP_2}(t)\}\right)$$
$$= P_1^{EGP}(t)z^{100} + P_2^{EGP}(t)z^{50} + P_3^{EGP}(t)z^0. \qquad (2.10)$$

where

$$P_1^{EGP}(t) = p_1^{EGP_1}(t) \cdot p_1^{EGP_2}(t)$$
$$P_2^{EGP}(t) = p_1^{EGP_1}(t) \cdot p_2^{EGP_2}(t) + p_2^{EGP_1}(t) \cdot p_1^{EGP_2}(t)$$
$$P_3^{EGP}(t) = p_2^{EGP_1}(t) \cdot p_2^{EGP_2}(t)$$

- Lz-transforms for *Conv*$_{2FP}$ subsystem:

$$L_z\{G^{Conv_{2FP}}(t)\} = \Omega_{f_{ser}}\left(L_z\{G^{TP}(t)\}, L_z\{G^{EGP}(t)\}\right)$$
$$= P_1^{Conv_{2FP}}(t)z^{100} + P_2^{Conv_{2FP}}(t)z^{50} + P_3^{Conv_{2FP}}(t)z^0. \quad (2.11)$$

where

$$P_1^{Conv_{2FP}}(t) = P_1^{TP}(t) \cdot P_1^{EGP}(t) + P_1^{TP}(t) \cdot P_2^{EGP}(t)$$
$$P_2^{Conv_{2FP}}(t) = P_2^{TP}(t) \cdot P_1^{EGP}(t) + P_2^{TP}(t) \cdot P_2^{EGP}(t)$$
$$P_3^{Conv_{2FP}}(t) = P_1^{TP}(t) \cdot P_3^{EGP}(t) + P_2^{TP}(t) \cdot P_3^{EGP}(t) + P_3^{TP}(t) \cdot P_1^{EGP}(t)$$
$$+ P_3^{TP}(t) \cdot P_2^{EGP}(t) + P_3^{TP}(t) \cdot P_3^{EGP}(t)$$

2.3.2.2 Conventional Aircraft with Four Fuel Pumps

According to Fig. 2.3, we will present the whole system using the Lz-transform as follows:

$$L_z\{G^T(t)\} = \Omega_{f_{par}}\left(L_z\{g^{LT}(t)\}, L_z\{g^{RT}(t)\}\right)$$
$$L_z\{G^{FP_{14}}(t)\} = \Omega_{f_{par}}\left(L_z\{g^{FP_1}(t)\}, L_z\{g^{FP_2}(t)\}, L_z\{g^{FP_3}(t)\}, L_z\{g^{FP_4}(t)\}\right)$$
$$L_z\{G^{TFP_{14}}(t)\} = \Omega_{f_{ser}}\left(L_z\{G^T(t)\}, L_z\{G^{FP_{14}}(t)\}\right),$$
$$L_z\{G^{EGP_1}(t)\} = \Omega_{f_{ser}}\left(L_z\{g^{E_1}(t)\}, L_z\{g^{G_1}(t)\}, L_z\{g^{P_1}(t)\}\right),$$
$$L_z\{G^{EGP_2}(t)\} = \Omega_{f_{ser}}\left(L_z\{g^{E_2}(t)\}, L_z\{g^{G_2}(t)\}, L_z\{g^{P_2}(t)\}\right),$$
$$L_z\{G^{EGP}(t)\} = \Omega_{f_{par}}\left(L_z\{G^{EGP_1}(t)\}, L_z\{G^{EGP_2}(t)\}\right)$$
$$L_z\{G^{Conv_{4FP}}(t)\} = \Omega_{f_{ser}}\left(L_z\{G^{TFP_{14}}(t)\}, L_z\{G^{EGP}(t)\}\right). \qquad (2.12)$$

For the expressions of Lz-transforms for tank, engines, gearboxes, and propellers subsystems, one can look in the Sect. 2.3.2.1. Using the composition operators Ω_{fpar}, we obtain the following Lz-transforms:

- Lz-transforms for FP_{14} subsystem:

$$
\begin{aligned}
L_z\{G^{FP_{14}}(t)\} &= \Omega_{f_{par}}\left(L_z\{g^{FP_1}(t)\}, L_z\{g^{FP_2}(t)\}, L_z\{g^{FP_3}(t)\}, L_z\{g^{FP_4}(t)\}\right) \\
&= \Omega_{f_{par}}\left(L_z\{G^{FP_{12}}(t)\}, L_z\{G^{FP_{34}}(t)\}\right) \\
&= \Omega_{f_{par}}\left(P_1^{FP_{12}}(t)z^{50} + P_2^{FP_{12}}(t)z^{25} + P_3^{FP_{12}}(t)z^0,\right. \\
&\quad\left. P_1^{FP_{34}}(t)z^{50} + P_2^{FP_{34}}(t)z^{25} + P_3^{FP_{34}}(t)z^0\right) \\
&= P_1^{FP_{14}}(t)z^{100} + P_2^{FP_{14}}(t)z^{75} + P_3^{FP_{14}}(t)z^{50} + P_4^{FP_{14}}(t)z^{25} \\
&\quad + P_5^{FP_{14}}(t)z^0.
\end{aligned}
\tag{2.13}
$$

where

$$
\begin{aligned}
P_1^{FP_{12}}(t) &= p_1^{FP_1}(t) \cdot p_1^{FP_2}(t) \\
P_2^{FP_{12}}(t) &= p_1^{FP_1}(t) \cdot p_2^{FP_2}(t) + p_2^{FP_1}(t) \cdot p_1^{FP_2}(t) \\
P_3^{FP_{12}}(t) &= p_2^{FP_1}(t) \cdot p_1^{FP_2}(t)
\end{aligned}
$$

$$
\begin{aligned}
P_1^{FP_{34}}(t) &= p_1^{FP_3}(t) \cdot p_1^{FP_4}(t) \\
P_2^{FP_{34}}(t) &= p_1^{FP_3}(t) \cdot p_2^{FP_4}(t) + p_2^{FP_3}(t) \cdot p_1^{FP_4}(t) \\
P_3^{FP_{34}}(t) &= p_2^{FP_3}(t) \cdot p_1^{FP_4}(t)
\end{aligned}
$$

$$
\begin{aligned}
P_1^{FP_{14}}(t) &= P_1^{FP_{12}}(t) \cdot P_1^{FP_{34}}(t) \\
P_2^{FP_{14}}(t) &= P_1^{FP_{12}}(t) \cdot P_2^{FP_{34}}(t) + P_2^{FP_{12}}(t) \cdot P_1^{FP_{34}}(t) \\
P_3^{FP_{14}}(t) &= P_1^{FP_{12}}(t) \cdot P_3^{FP_{34}}(t) + P_2^{FP_{12}}(t) \cdot P_2^{FP_{34}}(t) + P_3^{FP_{12}}(t) \cdot P_1^{FP_{34}}(t) \\
P_4^{FP_{14}}(t) &= P_2^{FP_{12}}(t) \cdot P_3^{FP_{34}}(t) + P_3^{FP_{12}}(t) \cdot P_2^{FP_{34}}(t) \\
P_5^{FP_{14}}(t) &= P_3^{FP_{12}}(t) \cdot P_3^{FP_{34}}(t)
\end{aligned}
$$

- Lz-transforms for TFP_{14} subsystem:

$$
\begin{aligned}
L_z\{G^{TFP_{14}}(t)\} &= \Omega_{f_{ser}}\left(L_z\{G^T(t)\}, L_z\{G^{FP_{14}}(t)\}\right) \\
&= P_1^{TFP_{14}}(t)z^{100} + P_2^{TFP_{14}}(t)z^{75} + P_3^{TFP_{14}}(t)z^{50} \\
&\quad + P_2^{TFP_{14}}(t)z^{25} + P_3^{TFP_{14}}(t)z^0
\end{aligned}
\tag{2.14}
$$

where

$$P_1^{TFP_{14}}(t) = P_1^T(t) \cdot P_1^{FP_{14}}(t)$$

$$P_2^{TFP_{14}}(t) = P_1^T(t) \cdot P_2^{FP_{14}}(t)$$

$$P_3^{TFP_{14}}(t) = P_1^T(t) \cdot P_3^{FP_{14}}(t) + P_2^T(t) \cdot P_1^{FP_{14}}(t)$$
$$+ P_2^T(t) \cdot P_2^{FP_{14}}(t) + P_2^T(t) \cdot P_3^{FP_{14}}(t)$$

$$P_4^{TFP_{14}}(t) = P_1^T(t) \cdot P_4^{FP_{14}}(t) + P_2^T(t) \cdot P_4^{FP_{14}}(t)$$

$$P_5^{TFP_{14}}(t) = P_1^T(t) \cdot P_5^{FP_{14}}(t) + P_2^T(t) \cdot P_5^{FP_{14}}(t)+$$
$$P_3^T(t) \cdot P_1^{FP_{14}}(t) + P_3^T(t) \cdot P_2^{FP_{14}}(t)$$
$$+ P_3^T(t) \cdot P_3^{FP_{14}}(t) + P_3^T(t) \cdot P_4^{FP_{14}}(t) + P_3^T(t) \cdot P_5^{FP_{14}}(t)$$

- Lz-transforms for $Conv_{4FP}$ subsystem:

$$L_z\{G^{Conv_{4FP}}(t)\} = \Omega_{f_{ser}}\left(L_z\{G^{TFP_{14}}(t)\}, L_z\{G^{EGP}(t)\}\right)$$
$$= P_1^{Conv_{4FP}}(t)z^{100} + P_2^{Conv_{4FP}}(t)z^{75} + P_3^{Conv_{4FP}}(t)z^{50}$$
$$+ P_2^{Conv_{4FP}}(t)z^{25} + P_3^{Conv_{4FP}}(t)z^{0}. \tag{2.15}$$

where

$$P_1^{Conv_{4FP}}(t) = P_1^{TFP_{14}}(t) \cdot P_1^{EGP}(t)$$

$$P_2^{Conv_{4FP}}(t) = P_2^{TFP_{14}}(t) \cdot P_1^{EGP}(t)$$

$$P_3^{Conv_{4FP}}(t) = P_3^{TFP_{14}}(t) \cdot P_1^{EGP}(t) + P_1^{TFP_{14}}(t) \cdot P_2^{EGP}(t)$$
$$+ P_2^{TFP_{14}}(t) \cdot P_2^{EGP}(t) + P_3^{TFP_{14}}(t) \cdot P_2^{EGP}(t)$$

$$P_4^{Conv_{4FP}}(t) = P_4^{TFP_{14}}(t) \cdot P_1^{EGP}(t) + P_4^{TFP_{14}}(t) \cdot P_2^{EGP}(t)$$

$$P_5^{Conv_{4FP}}(t) = P_5^{TFP_{14}}(t) \cdot P_1^{EGP}(t) + P_5^{TFP_{14}}(t) \cdot P_2^{EGP}(t)$$
$$+ P_1^{TFP_{14}}(t) \cdot P_3^{EGP}(t) + P_2^{TFP_{14}}(t) \cdot P_3^{EGP}(t)$$
$$+ P_3^{TFP_{14}}(t) \cdot P_3^{EGP}(t) + P_4^{TFP_{14}}(t) \cdot P_3^{EGP}(t)$$
$$+ P_5^{TFP_{14}}(t) \cdot P_3^{EGP}(t)$$

2.3.2.3 Conventional Aircraft with Six Fuel Pumps

According to Fig. 2.4, we will present the whole system using the Lz-transform as follows:

$$L_z\{G^T(t)\} = \Omega_{f_{par}}\left(L_z\{g^{LT}(t)\}, L_z\{g^{RT}(t)\}\right)$$

$$L_z\{G^{FP_{16}}(t)\} = \Omega_{f_{par}}\begin{pmatrix} L_z\{g^{FP_1}(t)\}, L_z\{g^{FP_2}(t)\}, L_z\{g^{FP_3}(t)\}, \\ L_z\{g^{FP_4}(t)\}, L_z\{g^{FP_5}(t)\}, L_z\{g^{FP_6}(t)\} \end{pmatrix}$$

$$L_z\{G^{TFP_{16}}(t)\} = \Omega_{f_{ser}}\left(L_z\{G^T(t)\}, L_z\{G^{FP_{16}}(t)\}\right),$$
$$L_z\{G^{EGP_1}(t)\} = \Omega_{f_{ser}}\left(L_z\{g^{E_1}(t)\}, L_z\{g^{G_1}(t)\}, L_z\{g^{P_1}(t)\}\right),$$
$$L_z\{G^{EGP_2}(t)\} = \Omega_{f_{ser}}\left(L_z\{g^{E_2}(t)\}, L_z\{g^{G_2}(t)\}, L_z\{g^{P_2}(t)\}\right),$$
$$L_z\{G^{EGP}(t)\} = \Omega_{f_{par}}\left(L_z\{G^{EGP_1}(t)\}, L_z\{G^{EGP_2}(t)\}\right)$$
$$L_z\{G^{Conv6FP}(t)\} = \Omega_{f_{ser}}\left(L_z\{G^{TFP_{16}}(t)\}, L_z\{g^{EGP}(t)\}\right). \tag{2.16}$$

For the expressions of Lz-transforms for tank, engines, gearboxes, and propellers subsystems, one can look in the Sects. 2.3.2.1 and 2.3.2.2. Using the composition operators $\Omega_{f_{par}}$, we obtain the following Lz-transforms:

- Lz-transforms for FP_{16} subsystem:

$$L_z\{G^{FP_{16}}(t)\} = \Omega_{f_{par}}\left(L_z\{g^{FP_1}(t)\}, L_z\{g^{FP_2}(t)\}, L_z\{g^{FP_3}(t)\}, L_z\{g^{FP_4}(t)\},\right.$$
$$\left. L_z\{g^{FP_5}(t)\}, L_z\{g^{FP_6}(t)\}\right)$$
$$= \Omega_{f_{par}}\left(L_z\{G^{FP_{14}}(t)\}, L_z\{G^{FP_{56}}(t)\}\right)$$
$$= \Omega_{f_{par}}\left(P_1^{FP_{14}}(t)z^{100} + P_2^{FP_{14}}(t)z^{75} + P_3^{FP_{14}}(t)z^{50}\right.$$
$$+ P_4^{FP_{14}}(t)z^{25} + P_5^{FP_{14}}(t)z^0,$$
$$\left. P_1^{FP_{56}}(t)z^{50} + P_2^{FP_{56}}(t)z^{25} + P_3^{FP_{56}}(t)z^0\right)$$
$$= P_1^{FP_{16}}(t)z^{150} + P_2^{FP_{16}}(t)z^{125} + P_3^{FP_{16}}(t)z^{100} + P_4^{FP_{16}}(t)z^{75}$$
$$+ P_5^{FP_{16}}(t)z^{50} + P_6^{FP_{16}}(t)z^{25} + P_7^{FP_{16}}(t)z^0. \tag{2.17}$$

where

$$L_z\{G^{FP_{16}}(t)\} = \Omega_{f_{par}}\left(L_z\{g^{FP_1}(t)\}, L_z\{g^{FP_2}(t)\}, L_z\{g^{FP_3}(t)\}, L_z\{g^{FP_4}(t)\},\right.$$
$$\left. L_z\{g^{FP_5}(t)\}, L_z\{g^{FP_6}(t)\}\right)$$
$$= \Omega_{f_{par}}\left(L_z\{G^{FP_{14}}(t)\}, L_z\{G^{FP_{56}}(t)\}\right)$$
$$= \Omega_{f_{par}}\left(P_1^{FP_{14}}(t)z^{100} + P_2^{FP_{14}}(t)z^{75} + P_3^{FP_{14}}(t)z^{50}\right.$$
$$+ P_4^{FP_{14}}(t)z^{25} + P_5^{FP_{14}}(t)z^0,$$
$$\left. P_1^{FP_{56}}(t)z^{50} + P_2^{FP_{56}}(t)z^{25} + P_3^{FP_{56}}(t)z^0\right)$$
$$= P_1^{FP_{16}}(t)z^{150} + P_2^{FP_{16}}(t)z^{125} + P_3^{FP_{16}}(t)z^{100} + P_4^{FP_{16}}(t)z^{75}$$
$$+ P_5^{FP_{16}}(t)z^{50} + P_6^{FP_{16}}(t)z^{25} + P_7^{FP_{16}}(t)z^0.$$

- Lz-transforms for TFP_{16} subsystem:

$$L_z\{G^{TFP_{16}}(t)\} = \Omega_{f_{ser}}\left(L_z\{G^T(t)\}, L_z\{G^{FP_{16}}(t)\}\right)$$
$$= P_1^{TFP_{16}}(t)z^{100} + P_2^{TFP_{16}}(t)z^{75} + P_3^{TFP_{16}}(t)z^{50}$$

$$+ P_2^{TFP_{16}}(t)z^{25} + P_3^{TFP_{16}}(t)z^0 \tag{2.18}$$

where

$$P_1^{TFP_{16}}(t) = P_1^T(t) \cdot P_1^{FP_{16}}(t) + P_1^T(t) \cdot P_2^{FP_{16}}(t) + P_1^T(t) \cdot P_3^{FP_{16}}(t)$$
$$P_2^{TFP_{16}}(t) = P_1^T(t) \cdot P_4^{FP_{16}}(t)$$
$$P_3^{TFP_{16}}(t) = P_1^T(t) \cdot P_5^{FP_{16}}(t) + P_2^T(t) \cdot P_1^{FP_{16}}(t) + P_2^T(t) \cdot P_2^{FP_{16}}(t)$$
$$\quad + P_2^T(t) \cdot P_3^{FP_{16}}(t) + P_2^T(t) \cdot P_4^{FP_{16}}(t) + P_2^T(t) \cdot P_5^{FP_{16}}(t)$$
$$P_4^{TFP_{16}}(t) = P_1^T(t) \cdot P_6^{FP_{16}}(t) + P_2^T(t) \cdot P_7^{FP_{16}}(t)$$
$$P_5^{TFP_{16}}(t) = P_1^T(t) \cdot P_7^{FP_{16}}(t) + P_2^T(t) \cdot P_7^{FP_{16}}(t) + P_3^T(t) \cdot P_1^{FP_{16}}(t)$$
$$\quad + P_3^T(t) \cdot P_2^{FP_{16}}(t) + P_3^T(t) \cdot P_3^{FP_{16}}(t) + P_3^T(t) \cdot P_4^{FP_{16}}(t)$$
$$\quad + P_3^T(t) \cdot P_5^{FP_{16}}(t) + P_3^T(t) \cdot P_6^{FP_{16}}(t) + P_3^T(t) \cdot P_7^{FP_{16}}(t)$$

- Lz-transforms for $Conv_{6FP}$ subsystem:

$$L_z\{G^{Conv_{6FP}}(t)\} = \Omega_{f_{ser}}\left(L_z\{G^{TFP_{16}}(t)\}, L_z\{G^{EGP}(t)\}\right)$$
$$= P_1^{Conv_{6FP}}(t)z^{100} + P_2^{Conv_{6FP}}(t)z^{75} + P_3^{Conv_{6FP}}(t)z^{50}$$
$$+ P_2^{Conv_{6FP}}(t)z^{25} + P_3^{Conv_{6FP}}(t)z^0. \tag{2.19}$$

where

$$P_1^{Conv_{6FP}}(t) = P_1^{TFP_{16}}(t) \cdot P_1^{EGP}(t)$$
$$P_2^{Conv_{6FP}}(t) = P_2^{TFP_{16}}(t) \cdot P_1^{EGP}(t)$$
$$P_3^{Conv_{6FP}}(t) = P_3^{TFP_{16}}(t) \cdot P_1^{EGP}(t) + P_1^{TFP_{16}}(t) \cdot P_2^{EGP}(t) + P_2^{TFP_{16}}(t) \cdot P_2^{EGP}(t)$$
$$\quad + P_3^{TFP_{16}}(t) \cdot P_2^{EGP}(t)$$
$$P_4^{Conv_{6FP}}(t) = P_4^{TFP_{16}}(t) \cdot P_1^{EGP}(t) + P_4^{TFP_{16}}(t) \cdot P_2^{EGP}(t)$$
$$P_5^{Conv_{6FP}}(t) = P_5^{TFP_{16}}(t) \cdot P_1^{EGP}(t) + P_5^{TFP_{16}}(t) \cdot P_2^{EGP}(t) + P_1^{TFP_{16}}(t) \cdot P_3^{EGP}(t)$$
$$\quad + P_2^{TFP_{16}}(t) \cdot P_3^{EGP}(t) + P_3^{TFP_{16}}(t) \cdot P_3^{EGP}(t) + P_4^{TFP_{16}}(t) \cdot P_3^{EGP}(t)$$
$$\quad + P_5^{TFP_{16}}(t) \cdot P_3^{EGP}(t)$$

2.3.2.4   Calculation of the Availability of the Conventional Aircraft

The instantaneous availability for constant demand level w may be presented as follows:

- For demand $w = 100\%$ and demand $w = 50\%$

Table 2.3 Failure and repair rates of elements in DH8D (year^{-1})

System's elements	Failure rates	Repair rates
Left and right tanks	0.06	75
Fuel pumps	0.25	120
Engines	0.6	45
Gearboxes	0.09	110
Propellers	0.06	95

$$A_{Conv2FP}(t) = \sum_{g_i^{Conv2FP}=100} P_i^{Conv2FP}(t) = P_1^{Conv2FP}(t).$$

$$A_{Conv4FP}(t) = \sum_{g_i^{Conv4FP}=100} P_i^{Conv4FP}(t) = P_1^{Conv4FP}(t).$$

$$A_{Conv6FP}(t) = \sum_{g_i^{Conv6FP}=100} P_i^{Conv6FP}(t) = P_1^{Conv6FP}(t). \qquad (2.20)$$

$$A_{Conv2FP}(t) = \sum_{g_i^{Conv2FP} \geq 50} P_i^{Conv2FP}(t) = P_1^{Conv2FP}(t) + P_2^{Conv2FP}(t).$$

$$A_{Conv4FP}(t) = \sum_{g_i^{Conv4FP} \geq 50} P_i^{Conv4FP}(t) = P_1^{Conv4FP}(t) + P_2^{Conv4FP}(t) + P_3^{Conv4FP}(t).$$

$$A_{Conv6FP}(t) = \sum_{g_i^{Conv6FP} \geq 50} P_i^{Conv6FP}(t) = P_1^{Conv6FP}(t) + P_2^{Conv6FP}(t) + P_3^{Conv6FP}(t).$$

$$(2.21)$$

The failure and repair rates (per year^{-1}) for each system's elements are presented in Table 2.3.

The calculated availability of conventional aircraft with different amount of fuel pumps after a year of operations is presented in Figs. 2.6 and 2.7.

As one can see, the instantaneous availability of the conventional aircraft is varying with the number of fuel pumps and depends on demand level. On the one hand, the analysis considers the cases using two and four pumps, where each pump, respectively, covers 50 and 25% of the needed system performance; hence, no redundancy option is provided. On the other hand, redundancy is provided for the case using six pumps, where each pump takes 25% of the needed system performance. The instantaneous availability of the conventional aircraft with 100% demand with six pumps is the greatest and with four pumps is the lowest. The instantaneous availability with 50% demand for all three variants is the same and is equal to 99.78%.

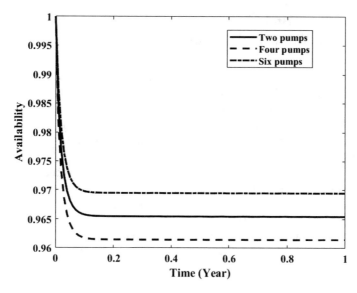

Fig. 2.6   Availability of conventional aircraft with varying number of fuel pumps with 100% demand

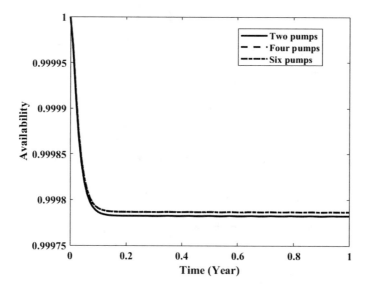

Fig. 2.7   Availability of conventional aircraft with varying number of fuel pumps with 50% demand

2.3.3 Lz-Transform Calculation of the Multi-state Models for the Electrified Aircraft

Even though the methodology was previously only explained for the conventional system, within this section the results are shown for both the conventional and electrical system, where two versions are assessed for the electrical system. The reliability block diagrams for the electrified system are shown in Figs. 2.8 and 2.9. The corresponding failure and repair rates of the elements used for the calculation of the availability are summarized in Table 2.4.

Using Lz-transform method, described in Sect. 2.3, the instantaneous availability of the electrified aircraft is varying with number of three- and nine-phase machines and depends on demand level (see Fig. 2.10). The instantaneous availability with 100% demand of the electrified aircraft with nine-phase motor is greater than the instantaneous availability of the electrified aircraft with three-phase motor on 0.99%. The instantaneous availability with 50% demand for both variants of the electrified aircraft is the same and is equal to 99.75%.

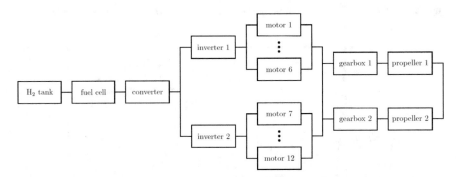

Fig. 2.8 Reliability block diagram of the electrified aircraft with 12 three-phase motors

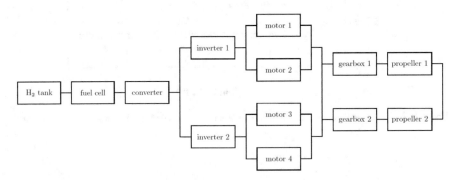

Fig. 2.9 Reliability block diagram of the conventional aircraft with four nine-phase motors

Table 2.4 Failure and repair rates of elements in DH8D (year^{-1})

System's elements	Failure rates	Repair rates
Hydrogen tanks	0.05	48
Fuel cells (without/with redundancy)	0.3 / 0.07	250 / 432
DC/DC converter	0.15	655
Inverter	0.2	584
Three-phase electric motor	0.09	113
Nine-phase electric motor	0.21	165
Propeller	0.06	95

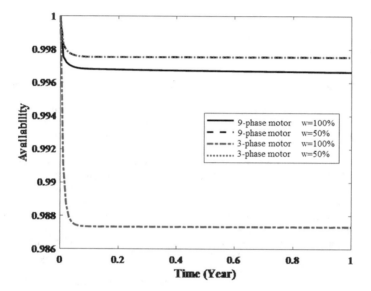

Fig. 2.10 Availability of electrified aircraft with varying number of three- and nine-phase motors

The slightly lower availability of the electric system—compared to the conventional one—results from the missing redundancy path for the electrical power supply subsystem.

2.4 Conclusion

Overall, the presented analysis shows that the range of the modified aircraft will not result in a beneficial range. The feasibility of fully electrifying the conventional aircraft is even less likely once the capacitors are factored in for other flight phases. Additionally, the replacement of the conventional propulsion system to an electric propulsion system in the studied model does not allow for the payload capacity to carry as many passengers as a conventional aircraft; as a result, more flights are necessary or more airplanes are needed in order to meet the demands of air travel. Still, the analysis does not factor in the weight reduction of the replacing hydraulic, pneumatic, or other non-propulsive systems with an electrical system, which may allow for additional fuel capacity. Furthermore, the availability assessment shows that the electric version is close to the values of the conventional one, even without redundancy of the supply subsystem. The increase of the number of fuel pumps in the conventional system in order to produce redundancy does not have a significant impact. In future publications, the redundancy for the electrified system could be included as well as different options of topologies.

References

1. R. Berger, Aircraft electrical propulsion-onwards and upwards. Think Act. Roland Berger LTD (2018)
2. I. Bolvashenkov, I. Frenkel, J. Kammermann et al., Comparison of the battery energy storage and fuel cell energy source for the safety-critical drives considering reliability and fault tolerance. In *Proc. of IEEE International Conference on Information and Digital Technologies (IDT)*, ˇ Zilina, Slovakia , 5–7 July 2017, pp. 63–70
3. I. Bolvashenkov, H.G. Herzog, F. Ismagilov et al., *Fault-Tolerant Traction Electric Drives: Reliability, Topologies and Components Design* (Springer, Singapore, 2020)
4. I. Bolvashenkov, J. Kammermann, Q. Buchner et al., A traction drive of an electrical helicopter based on fuel cells and superconducting electrical machines: preliminary assessment of feasibility. In *Proc. of 7ʰ International Conference & Workshop Energy For Tomorrow (REMOO'17)*, Venice, Italy , 10–12 May 2017, pp. 1–10
5. I. Bolvashenkov, J. Kammermann, H.G. Herzog, Electrification of helicopter: actual feasibility and prospects. In *Proc. of 13th IEEE Vehicle Power and Propulsion Conference (VPPC'17)*, Belfort, France, 11–14 December 2017, pp. 1–6
6. G. Evans, Advanced flight simulation fuel planning. (Fuelplanner, 2020). http://fuelplanner.com/. Accessed 15 March 2020
7. M. Filipenko, S. Biser, M. Boll et al., Optimization of turboelectric propulsion systems for short range passenger aircraft. (Aerospace, 2020), p. 23
8. M. Hepperle, Electric flight-potential and limitations. In *AVT-209 Workshop on Energy Efficient Technologies and Concepts Operation*, Lisbon, Portugal, 22–24 October 2012
9. J. Kammermann, Potential analysis of electrical drive trains according to application requirements. Dissertation, Department of Electrical and Computer Engineering, Technical University of Munich (TUM), Munich, Germany, 2019. https://doi.org/10.14459/2019md1451565
10. J. Kammermann, I. Bolvashenkov, K. Tran, H.G. Herzog, I. Frenkel, Feasibility study for a full-electric aircraft considering weight, volume, and reliability requirements. In *International Conference on Electrotechnical Complexes and Systems (ICOECS)*, (IEEE, Ufa, Russia, 2020)
11. A. Lisnianski, I. Frenkel, Y. Ding, *Multi-State System Reliability Analysis and Optimization for Engineers and Industrial Managers* (Springer, London, 2010)
12. A. Lisnianski, I. Frenkel, L. Khvatskin, *Modern Dynamic Reliability Analysis for Multi-state Systems*. Springer Series in Reliability Engineering (Springer, Cham, 2021)
13. C. Pornet, *Electric Drives for Propulsion System of Transport Aircraft New Applications of Electric Drives* (IntechOpen, London, 2015), pp. 115–141
14. S. Sahoo, X. Zhao, K. Kyprianidis, A review of concepts, benefits, and challenges for future electrical propulsion-based aircraft. Aerospace **7**(4), 44 (2020)

15. D. Scholz, Evaluating aircraft with electric and hybrid propulsion. In *UKIP Media & Events: Conference Proceedings: Electric & Hybrid Aerospace Symposium,* (Cologne, Germany. November 2018)

16. J. Wilkerson, M. Jacobson, A. Malwitz et al., Analysis of emission data from global commercial aviation: 2004 and 2006. Atmos. Chem. Phys. **10**(13), 6391–6408 (2010)

Chapter 3
Design and Feasibility of Electrical Version of Search-and-Rescue Helicopter Based on Eurocopter

3.1 Introduction

The electrification of various types of air transport is now a very urgent problem, due to a number of well-known advantages compared to conventional types. One important area is the electrification of helicopters. With the constant growth of complexity within modern electric engineering systems, it complicates the task of achieving the required level of its sustainable and safe operations taking into account the satisfaction of stringent requirements for the weight and overall dimensions of the propulsion system, to ensure the specified flight range (Fig. 3.1). Such important features of an electric propulsion system for helicopters have not been sufficiently investigated up to now, due to the lack of compact, reliable electric energy sources with high-energy density. Feasibility of realization electric version of helicopter's propulsion system depends entirely on the optimal selection and placement of each component of the traction drive. It should be noted that the analysis of cost characteristics has not been considered in this chapter.

The constant increase in volume of air traffic makes especially the task urgent how to provide more sustainable flying with a minimum of energy consumption and without further environmental damage. That is why the application of an electrical propulsion system for the search-and-rescue helicopter has caused the great interest now. The development of "electric" aircraft requires a comprehensive revision of the design principles of a variety of devices and systems of the aircraft, which is associated with the creation of the new electric traction drive having a low specific weight and consisting of electric energy sources, electric energy converters, and electric traction motors as shown in Fig. 3.2. The use of electrical technologies could lead in the near future to a change in the principles of its construction.

The electric energy source is one of the main and important parts of the full-electric vehicle's traction drive. The safety of helicopter's flight mainly depends on the sustainability of onboard electric energy sources. Today, there are really many firms and research institutes all-around the world, which are all dedicated

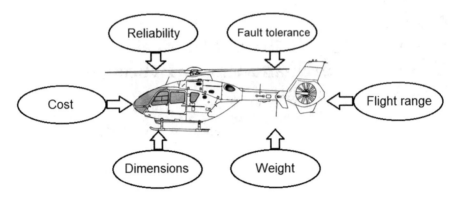

Fig. 3.1 Specific features of helicopter's design

Fig. 3.2 Structure of
helicopter's electric
propulsion system

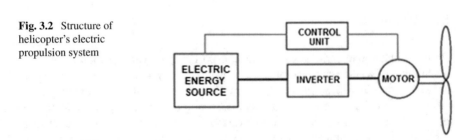

to the development and optimization of the battery electric energy storage (BEES) performance.

At the same time, due to the fact that the batteries at the modern level of technological development have insufficient values of energy and power density in comparison with hydrogen energy carrier, as is evident from Fig. 3.3, it is expedient to carry out a comparative analysis of electric drive options with other competitive electric energy sources. As such an alternative option for comparison, the fuel cell electric energy source (FCEES) has been chosen.

Considering the purpose of using the battery electric energy storage, the accurate assessment of indicators, such as energy density, power density, storage capacity, failure probability, and fault tolerance of FCEES, is becoming extremely important for the optimal choice of topology and the parameters of such supply devices. In case of an electric helicopter, the battery cell's or fuel cell's failures can lead to a reduced functionality, deterioration of operational modes, and even a catastrophic situation in the electric propulsion system.

Due to the specifics of the operational modes of helicopters in contrast to airplanes, the traction drive requires more power and higher-energy density, which in turn leads to the need of a more powerful electric traction motor and a larger capacity of electric energy source or energy storage.

This is limited by the maximum takeoff weight (MTOW) of the helicopter and the space for installation of the electric traction drive components. For this reason,

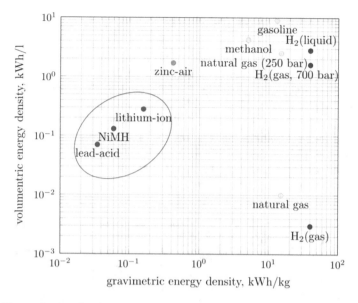

Fig. 3.3 Energy density of various storage systems [1]

as shown in [2, 8], the use of a battery energy storage as a source of electrical energy is currently not feasible because of insufficient energy density and large weight and dimensions.

Despite the large amount of research, superconducting electric motors do not find practical use for the traction drives of aircraft. Similar trends are also observed for the use of traction electric drives with fuel cells and conventional electric motors for aircraft electrification [37]. This is because the maximum synergy effect can be obtained only by applying the fuel cells together with different variants of superconducting electrical machines. Therefore, the present study attempts to assess the possibility of using fuel cells and superconducting motors to create a small electrical helicopter.

Today, it has become topical, because in the last five years the great technological breakthroughs regarding the development of superconducting motors [13, 15, 16, 23, 25–27] and modern fuel cells with significantly improved efficiency, weight, dimensions, and cost characteristics [17, 18, 30, 35, 37] have been realized. Based on the analysis of the collected contemporary data, the implementation of an all-electric version of the helicopter was considered to be feasible.

The conceptual system, that will electrify the helicopter, consists of the following subsystems: traction electric motor, proton exchange membrane fuel cells, power electronics, and hydrogen tank. The parameters of these subsystems are examined with regard to their impact on the entire systems mass, volume, and fault tolerance. The results of this analysis on the possibility of implementing a search-and-rescue electric helicopter of the small class today and tomorrow are presented in this chapter.

3.2   Conventional Object of Study

As a basic prototype for the comparative analysis and subsequent modernization in a purely electric version, the airbus helicopter EC135 currently being in operation has been considered, as shown in Fig. 3.4. The traction drive of the EC135 has two gas turbine engines and a speed reducer and is described in more detail in [31]. Accordingly, the turbines Turbomeca Arrius-2B2 or Pratt and Whitney PW206B2 are installed as gas turbine engines on the EC135. Technical and design data of a conventional EC135 are shown in Table 3.1.

According to the statistics from the German automobile club (ADAC), every SAR helicopter in Germany is operated by a daily average of 8–10 h, i.e., the average ratio

Fig. 3.4   Helicopter EC135 [31]

Table 3.1   Technical data of propulsion system

Technical data	Value
Maximal power per engine	Arrius-452 kW, P&W-463 kW
Fuel consumption	232 l/h
Mean range value	615 km
Maximal speed	259 km/h (140 knots)
Service ceiling	3045 m (10,000 feet)
Empty weight	1490 kg
Maximal takeoff weight	2980 kg
Cabin volume with place for pilot	3.8 and 1 cu m
Baggage compartment volume	1.1 cu m

Table 3.2 Technical data of EC135 traction drive

Component	Weight, kg	Volume, l
Two turbine engines	228	330
Fuel tank	650	737
Total	878	1067

of operating time in one year is 0.33–0.42. Thus, in the future simulation, an annual flight of the helicopter is assumed to be equal to 3000 h.

Table 3.2 shows the weight and dimensions of a traditional traction drive of the EC135 with two turbines. For the electrification of the propulsion system of the helicopter, these values have been taken as a given project restriction.

3.3 Electric Version of Traction Drive

In this section, the main parameters of the basic components of an electric traction drive of a helicopter are considered: weight, volume, and fault tolerance. For further calculations, the gravimetric and volumetric densities of the traction electric drive components were determined on the basis of modern data of electric motors, power electronics, fuel cells, and fuel tanks that are already in operation or of the corresponding experimental laboratory samples.

Taking into account the specifics of the operation of the rescue helicopter, the main types of flights performed by it throughout the entire life cycle are presented in Fig. 3.5. These special flights are search-and-rescue (SAR), medical evacuation (MEDEVAC), and emergency medical services (EMS).

Different operating modes correspond to different rescue missions of the helicopter: takeoff, climb, cruise, descent, hovering, and landing. Each of these modes is characterized by the corresponding value of the required power of the traction electric drive.

A comparative analysis of the basic characteristics of the traction electric drive components was made on the basis of the required power of the rescue helicopter throughout the entire flight, which is shown graphically in Fig. 3.6 for a one-hour flight.

Comprehensive reliability indicators of individual components and the electric drive as a whole were calculated by the authors on the basis of stochastic models, described in [3, 5, 6, 8, 20, 21]. The calculation of the reliability indicators was

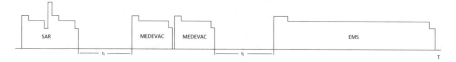

Fig. 3.5 Operating cycles of the helicopter during lifetime

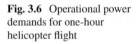

Fig. 3.6 Operational power demands for one-hour helicopter flight

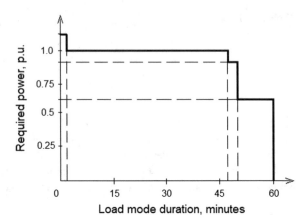

carried out in order to analyze the compliance with the requirements for the fault tolerance of the helicopter propulsion system. The probability of complete failure of the helicopters traction electric drive should not exceed 10^{-9}/h.

3.3.1 Traction Electric Motor

As shown by previous studies [12], a synchronous motor with permanent magnets is currently optimal for use as a traction motor for a helicopter from the point of view of a systematic approach.

3.3.1.1 Weight and Volume

The future electrification of the air transport imposes extremely stringent requirements on the weight and dimensions of the used electrical machines. Paying attention to the prospects for use for air transport, a superconducting version of an electric motor was considered.

Today, a lot of research is carried out in the field of different types of superconducting machines. More information is discussed in [13, 15, 16, 19, 23, 25–28, 33]. One promising design for 1 MW high-temperature superconducting motor, described in [23], is shown in Fig. 3.7.

Despite the promising results of investigations in the Ohio State University and University of Illinois [26] on the superconducting motors with a power density of 13 kW/kg, the present analysis was carried out by taking more modest values as characteristics of the existing superconducting motors, i.e., a gravimetric power density of 7.6 kW/kg and a volumetric power density of 13.8 kW/l, respectively [28].

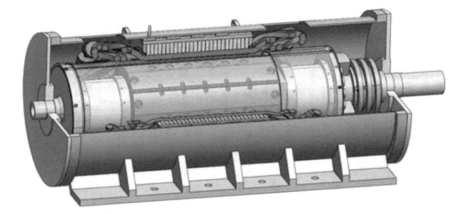

Fig. 3.7 High-temperature superconducting motor [23]

3.3.1.2 Reliability and Fault Tolerance

As shown in [12], a synchronous motor with permanent magnets (PMSM) and concentrated stator windings is currently optimal for use as a traction motor for a helicopter in considering of a systematic approach. From the point of view of the fault tolerance of an electrical machine, the most important design characteristic is the number of motor phases.

In accordance with requirements as well as statistical data on the reliability of multi-phase traction electric motors, it was identified that the suitable model for analysis of the fault tolerance is a multi-state system Markov model (MSS MM) with a minimum of four states, as shown in Fig. 3.8. The main reason for this choice is explained by the fact that with only one degradative state of traction electric motor, it is not feasible to realize the required values of fault tolerance.

The values of the transition probabilities λ are calculated based on the thermal calculation results of the electric motor for failure conditions, considering the overload capacity and thermal stability of the motor, especially of its stator windings. The calculation is performed based on the degree of fault tolerance, described in [4, 7].

Regarding the design requirements on fault tolerance of an electric helicopter, the total failure probability of the traction electric motor at reduced power of 65 or 85% of the nominal value, as well as at 115% of nominal value, was analyzed using

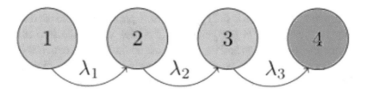

Fig. 3.8 Multi-state Markov model of traction multi-phase PMSM

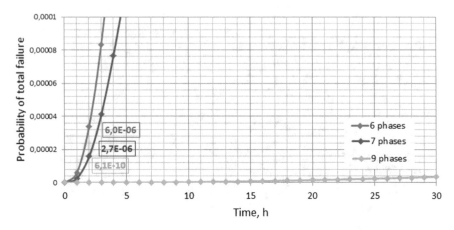

Fig. 3.9 Probability of total failure of electric motor at the 115% load

Table 3.3 Probabilities of complete failure of electric motor

Load level (%)	Phase number		
	Six phases	Seven phases	Nine phases
115	6.8×10^{-6}	3.7×10^{-6}	6.5×10^{-10}
85	8.4×10^{-9}	8.3×10^{-10}	5.2×10^{-12}
65	2.7×10^{-11}	2.7×10^{-12}	7.7×10^{-16}

the MSS MM. The corresponding graph for six-, seven-, and nine-phase PSM at the 115% load level is presented in Fig. 3.9.

Table 3.3 shows the values of the probability of a complete failure traction electric motor for the given load levels—65, 85, and 115% of the nominal value.

As can be seen from the graphs in Fig. 3.9 and Table 3.3, the simulation results of two consistent critically dangerous failures allow evaluating the fault tolerance value of a multi-phase electric motor, which is one important part in the traction drive of electric helicopters. The nine-phase PMSM have shown the maximum compliance with the requirements relating to the safety–critical drive.

3.3.2   Power Electronics

For the assessment of the electric energy converters parameters, the promising multi-level topology of cascaded H-bridge (CHB) inverter was taken into account and accepted, since the dimensions and the weight of the multi-level converter compared to the conventional topology have significantly smaller values, as discussed in [2]. On the other hand, from a fault tolerance point of view, the preferred topology is a 17-level CHB inverter based on MOSFETs as shown in [11].

Fig. 3.10 DC/DC converter [29]

Table 3.4 Technical data of semiconductor converters

Power electronics	Power density, kW/kg	Power density, kW/l
DC/DC converter	64	143
DC/AC converter	30	69

3.3.2.1 Weight and Volume

An experimental electric converter sample of the Fraunhofer Institute (Germany) has been used [29] for design analysis of the DC/DC and DC/AC converter since they have high-power density compared to other converters and were designed especially for the operation with a fuel cell systems. One of these components is shown in Fig. 3.10.

The values of gravimetric and volumetric power densities resulting from the calculation of the parameters of the mentioned experimental samples are shown in Table 3.4.

3.3.2.2 Reliability and Fault Tolerance

Considering the technical design parameters of the electric helicopter, the preferred option is a topology of 17-level CHB inverter, including the features and advantages which are well represented in [11]. Thus, in each phase there are eight inverter submodules.

Based on the overload capability of multi-level inverter, described in [7], and real temperature modes during overload of an inverter in the case of successive failures of one, two, or more inverter submodules, the transition probabilities have been calculated for the MSS MM, similar to the calculations for electric motors, as shown in Fig. 3.11.

Figure 3.12 is a graphical representation of simulation results on MSS MM of the probability of a total failure of an electric 17-level inverter at increased power up to 115% of the nominal value for the six-, seven-, and nine-phase versions of traction motor.

Table 3.5 shows the values of the probability of a complete failure multi-level inverter for required load levels—65, 85, and 115% of the nominal value.

The simulation results of three consecutive critically dangerous failures in the same phase allow quantifying the degree of fault tolerance of a 17-level inverter, which is one of the important parts of the helicopter's traction drive. The nine-phase option has shown the maximum compliance with the requirements relating to the safety–critical drives.

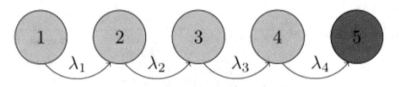

Fig. 3.11 Multi-state Markov model of 17-level inverter

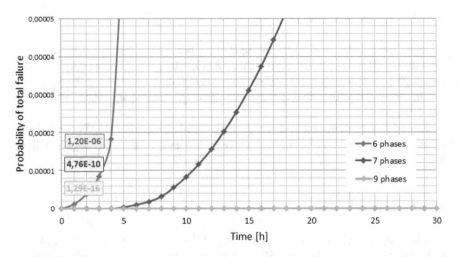

Fig. 3.12 Probability of a total failure of one motor phase with 17-level inverter at the 115% load

Table 3.5 Probabilities of complete failure of inverter

Load level (%)	Phase number		
	Six phases	Seven phases	Nine phases
115	1.2×10^{-6}	4.8×10^{-10}	1.3×10^{-16}
85	1.8×10^{-7}	7.7×10^{-11}	1.7×10^{-18}
65	2.7×10^{-8}	7.5×10^{-13}	6.7×10^{-20}

3.3.3 Electric Energy Source and Hydrogen Storage

As noted above, the current level of development of electric batteries does not allow even a one-hour flight of an electric helicopter to be realized. Therefore, in this chapter, FCEES, which are fuel cells, hydrogen tanks, and ultracapacitors, were considered as a source of electric energy for the helicopter.

As is well known, the FCEES consist of multiple cells in parallel and/or serial connections in order to satisfy the high power and reliability design requirements to the traction drive of electrical helicopters. At the same time, it is necessary to consider the strict limitations on the installation space and weight of the whole electric propulsion system. The entire system reliability assessment is based on the reliability evaluation of system components including individual cells and stacks.

3.3.3.1 Weight and Volume

Just a few years ago, the fuel cells were significantly losing against an electrical traction drive with battery electric energy storage in almost all parameters [17]. However, nowadays, fuel cells are at least not inferior to them regarding all main characteristics [18, 22, 24, 30, 35, 38]. Nevertheless, for further assessment the real fuel cell's data, installed on the Toyota Mirai, were taken [38], as shown in Table 3.6.

It should be noted that in the near future in accordance with the project AutoStack-CORE, described in [22], it is planned to create a fuel cell with higher performance in comparison with the fuel cell of Toyota.

The design of recent option of the fuel cell is presented in Fig. 3.13.

For the correct assessment of the weight parameters of FCEES, the inertia of the fuel cells was taken into account, as shown in Fig. 3.14. One possible way to improve the dynamic characteristics of FCEES is to use ultracapacitors (UC), as shown in [32, 36]. Thus, the weight of the required number of UC was taken into account in order to evaluate the total weight of the FCEES. Figure 3.15 shows the characteristics of UC in comparison with other electrical energy storage devices.

Table 3.6 Technical data of fuel cells

Fuel cell type	Power density, kW/kg	Power density, kW/l
Toyota Mirai	2.0	3.1
AutoStack	2.8	3.4

Fig. 3.13 Fuel cell design [24]

Fig. 3.14 Power response of FCs, batteries, and UCs [36]

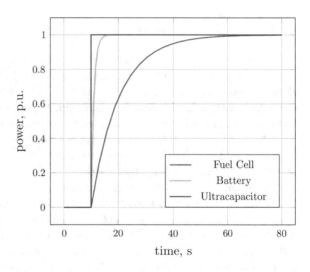

In the FCEES, hydrogen serves as the energy carrier and is used to provide the necessary amount of energy for the traction drive of electrical helicopter. Based on the analysis of modern technologies for hydrogen storage, two of the most suitable configurations for hydrogen storage tanks were chosen: a cylindrical tank [14] and a spherical tank [34]. The technical specifications of the considered variants for the further assessment are shown in Table 3.7.

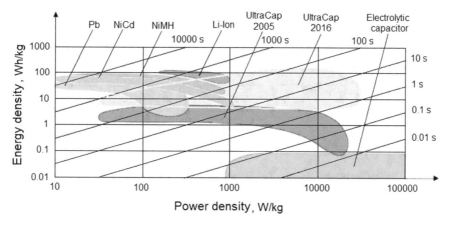

Fig. 3.15 Energy and power density of electric energy storages [1]

Table 3.7 Technical data of hydrogen storage

Storage tank	Energy density, kWh/kg	Energy density, kWh/l
Cylindrical (BMW), 40 kg gas H_2	3.0	1.2
Spherical, 60 kg liquid H_2	5.8	1.6

Table 3.8 Total weight of FCEES in kilogram

Redundancy	Component			
	FC	Tank	UC	Total
Without	338	195	167	700
With	406	195	201	802

As hydrogen storage, the characteristic of the spherical variant of tank has been chosen for the calculation. Table 3.8 shows the weight characteristics of the particular components and the entire FCEES.

It is theoretically possible and technically expedient to arrange multiple hydrogen tanks outside of the helicopter, but in the present chapter this opportunity of the placement of hydrogen tanks was not considered.

3.3.3.2 Reliability and Fault Tolerance

Considering the characteristics of conventional helicopter, as the parameters of the electric traction drive were accepted: the output nominal power is 500 kW, the traction motor rotational speed is 400 rpm, and the voltage is 800 V. In accordance with the above parameters, the minimally sufficient number of fuel cells and their connections

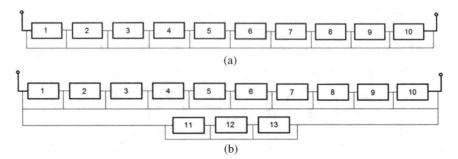

(a)

(b)

Fig. 3.16 Reliability block diagram of fuel cell electric energy source, **a** without redundancy, **b** with 30% redundancy

schemes in terms of the maximum fault tolerance has been determined. The schemes are shown in Fig. 3.16.

According to the method described in [1, 2, 9, 10], each fuel cell stack can be represented as a device with two states of performance: a fully operational state with a capacity G (50 kW) and a total failure corresponding to a capacity of 0, as shown in Fig. 3.17.

According to the Markov method and the state transition diagram, the following system of differential equations was built:

$$\begin{cases} \dfrac{dp_1(t)}{dt} = -\lambda p_1(t), \\ \dfrac{dp_2(t)}{dt} = \lambda p_2(t). \end{cases}$$

Initial conditions are as follows: $p_1(0) = 1$, $p_2(0) = 0$.

A numerical solution for probabilities can be obtained for each of this system of differential equations using MATLAB®. Therefore, one obtains for each stack the following output performance stochastic processes:

$$\mathbf{g} \in \{g_1, g_2\} = \{1, 0\};$$
$$\mathbf{p}(t) = \{p_1(t), p_2(t)\}.$$

Having the set one can define L_z-transform, associated with each accumulator output performance stochastic process:

$$L_z\{g(t)\} = p_1(t)z^1 + p_2(t)z^0.$$

Taking the assumption that all fuel cell stacks are similar, the final expression of the whole system's L_z-transform is of the following form:

Fig. 3.17 State transitions
diagram for fuel cell electric
energy source

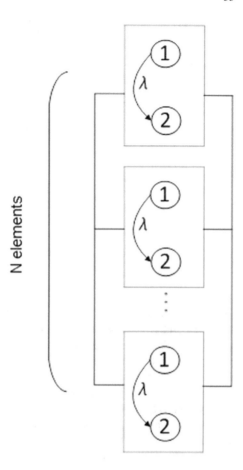

$$L_z\{G(t)\} = \prod_{i=1}^{n} \{p_1(t)z^1 + p_2(t)z^0\}$$

$$= \sum_{k=0}^{n} \binom{n}{k} \{p_1(t)z^1\}^{n-k} \{p_2(t)z^0\}^k$$

Using the whole system's L_z-transform is possible to obtain the multi-state system
(MSS) reliability function for the constant demand level. For the system without
redundancy, this function has the following form:

$$A_1(t) = \sum_{k=0}^{0} \binom{10}{k} p_1(t)^{10-k} p_2(t)^k = p_1(t)^{10}$$

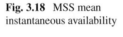

Fig. 3.18 MSS mean instantaneous availability

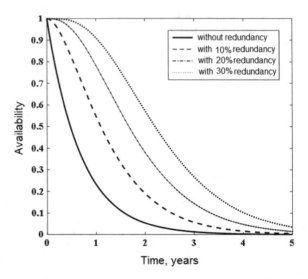

Despite the fact that currently, there are publications about the higher rates of reliability indices of the fuel cell stacks, for the calculations the minimum MTTF value of 20,000 h was accepted.

Figure 3.18 presents the simulation results of the MSS mean instantaneous availability for the fuel cells electric energy source with the different levels of redundancy: 0, 10%, 20%, and 30%, respectively. It should be noted that the reliability indices of hydrogen storage system in the calculations were not considered.

Based on the results of simulation, the level of redundancy of electric energy source, sufficient for the implementation of the required fault tolerance values, has been defined. The results have shown that the fault tolerance indices of the fuel cell stacks could be improved to the required value by implementing more than 20% redundant battery cells.

3.4   Evaluation of Feasibility

The above-discussed results have shown that the optimum in terms of the weight and volume of the embodiment of the FCEES submodules in considering of the fault tolerance requirements is a nine-phase version of a superconducting electric motor with a 17-level electric CHB inverter. Based on the obtained technical data of the main components of the propulsion system, an analysis was made of the opportunity of implementing an electric version of the helicopter for the one-hour flight. Table 3.9 presents the results of comparative evaluation of the different options.

The value of gravimetric density of 10, 30, 50 Wh/kg, and, respectively, volumetric density of 8, 30, and 60 Wh/l, achieved by the various researchers on the basis of graphene's UCs, as the basic values to calculate the weight and volume characteristics

Table 3.9 Technical data of electric version of helicopter propulsion system

Component	Weight, kg	Volume, l
Motor	69	38
DC/DC converter	12.8	5.6
DC/AC converter	17.4	9.6
Fuel cell, Toyota	338	217
Spherical tank	195	693
Cylindrical tank	347	866
Total (spherical tank)	632.2	963.2
Total (cylindrical tank)	784.2	1136.2

of ultracapacitors has been accepted. It improves the dynamic features of the fuel cells, and UCs should provide sustainably the electric traction drive of the helicopter with required energy for 60 s until the moment when it reaches the nominal mode of generating power.

Tables 3.10 and 3.11 show the weight and volume characteristics of ultracapacitors.

Based on the calculated data, two graphs were constructed, presented in Figs. 3.19 and 3.20. These two figures show the total weight and volume of traction electric drive as a function of the FC efficiency and UC energy density. As initial data, an energy density of spherical type of hydrogen tank as well as the power density of the fuel FC of Toyota Mirai type has been used. Looking at the overall weight and the overall volume of propulsion system, the energy conversion efficiency of the fuel cell stacks is the most important influencing factor. This is because of the strict depending from efficiency value required for the flight quantity of hydrogen and, respectively, the weight and volume of the hydrogen tank changes. Thus, the calculations of the feasibility of the one-hour flight were carried out based on the necessary energy amount, stored in the hydrogen tank in considering all energy transformation processes in traction drive.

Table 3.10 Total weight of UC in kilogram

Type of UC Redundancy	10 Wh/kg	30 Wh/kg	50 Wh/kg
Without	835	278	167
With	1002	334	201

Table 3.11 Total volume of UC in liter

Type of UC Redundancy	8 Wh/l	30 Wh/l	60 Wh/l
Without	1043	278	139
With	1251	334	167

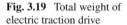

Fig. 3.19 Total weight of
electric traction drive

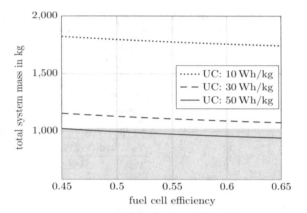

Fig. 3.20 Total volume of
electric traction drive

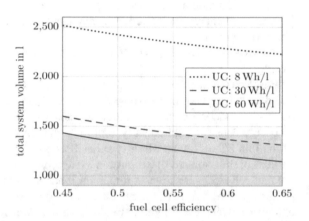

The green area on the graph shows how much free installation space (weight and volume) is extracted from the helicopter by the removal of two turbines with transmissions and the fuel tank with fuel. As can be seen from Fig. 3.19, the heaviest components of the electric drive are the fuel cells unit and a hydrogen tank. Their weights are determined by the calculation of total weight of the electric drive. It should be noted that overall system mass decreases by increasing of energy conversion efficiency. The characteristics of the total weight variations can be used successfully for the electric drive components with the more advanced initial data. Only the option with advanced ultracapacitors with energy density of 50 Wh/kg is in the green area of the graph for the whole range of the fuel cells efficiency. This means that with an actual level of the UC performance (10 Wh/kg), electric flight is not feasible.

Considering Fig. 3.20, it becomes clear that for an actual performance level of the FC and UC, electric traction drive is not feasible. But for the both variants of an advanced ultracapacitors, the curves are in the green area of the graph. This means that by using of such topology and components, an electric traction drive for the helicopters electrification is feasible. In order to increase the useful volume of the helicopter and the safety landing in autorotation mode in emergency case, it is advisable to investigate the use of an external arrangement of multiple hydrogen tanks with reduced volume. Thus, the number of tanks can be determined by the range value of the planned flight.

3.5 Conclusion

Based on the preliminary assessment of weight, volume, and fault tolerance of helicopter's electric traction drive, it was concluded that the advancements in superconducting electric machines technologies, fuel cells, and hydrogen tanks development have the sufficient potential to provide significant improvements of helicopter performance and the possibility of its practical realization.

The main condition for the successful implementation of this concept of electrical helicopter based on a conventional search-and-rescue helicopter type EC135 is the joint use of superconducting motors, cooled by the liquid hydrogen, and electric energy source on the basis of the fuel cells and ultracapacitors. Based on the availability simulation results, it has been concluded that the fuel cell stacks reliability could be improved to the required fault tolerance value by implementing of more than 20% cells redundancy.

With the further improvements of the characteristics of fuel cells, ultracapacitors, superconducting motors, hydrogen storage, and electric converters, it is advisable to update the results of these evaluation in accordance with the new advanced technical data of the components of the helicopter's electric propulsion system.

References

1. I. Bolvashenkov, I. Frenkel, J. Kammermann et al., Comparison of the battery energy storage and fuel cell energy source for the safety-critical drives considering reliability and fault tolerance, in *Proc. of IEEE Int. Conference on Information and Digital Technologies (IDT)*, Žilina, Slovakia , 5–7 July 2017, pp. 63–70
2. I. Bolvashenkov, I. Frenkel, J. Kammermann et al., The choice of an optimal structure and parameters of energy storage for an electrical helicopter traction drive, in *Proc. of 12th International Conference and Exhibition on Ecological Vehicles and Renewable Energies (EVER)*, Monaco, 11–13 April 2017, pp. 1–6
3. I. Bolvashenkov, H.G. Herzog, Use of stochastic models for operational efficiency analysis of multi power source traction drives, in *Proc. of the 2d Int. Symposium on Stochastic Models in Reliability Engineering, Life Science and Operations Management, (SMRLO)*, Beer Sheva, Israel, 15–18 February 2016, pp. 124–130
4. I. Bolvashenkov, H.G. Herzog, Degree of fault tolerance of the multi-phase traction electric motors: methodology and application. In *Proc. of 16th IEEE Int. Conf. on Environment and Electrical Engineering (EEEIC'16)*, Florence, Italy, 7–10 June 2016, pp. 1–6
5. I. Bolvashenkov, H.G. Herzog, I. Frenkel et al., *Safety-Critical Electrical Drives: Topologies, Reliability, Performance* (Springer, Cham, Switzerland, 2018)

6. I. Bolvashenkov, H.G. Herzog, F. Ismagilov et al., *Fault-Tolerant Traction Electric Drives: Reliability, Topologies and Components Design* (Springer, Singapore, 2020)
7. I. Bolvashenkov, J. Kammerman, H.G. Herzog, Methodology for Determining the Transition Probabilities for Multi-State System Markov Models of Fault Tolerant Electric Vehicles in *Proc. of Int. Asian Conference on Energy, Power and Transportation Electrification (ACEPT)*, Singapore, 25–27 October 2016, p. 6
8. I. Bolvashenkov, J. Kammerman, H.G. Herzog, Reliability assessment of a fault tolerant propulsion system for an electrical helicopter, in *Proc. of 12th International Conference and Exhibition on Ecological Vehicles and Renewable Energies (EVER)*, Monaco, 11–13 April 2017, pp. 1–6
9. I. Bolvashenkov, J. Kammerman, H.G. Herzog, Investigation of reliability and fault tolerance of multiphase traction electric motor supplied with multi power source based on Lz-transform, in *Proc. of IEEE Int. Conference on System Reliability and Safety (ICSRS)*, Milano, Italy, 20–22 December 2017, pp. 303–309
10. I. Bolvashenkov, J. Kammerman, H.G. Herzog, Fault tolerant traction drive of electrical helicopter with battery electric energy storage. in *Proc. of 17th IEEE Int. Conference on Environment and Electrical Engineering (EEEIC)*, Milano, Italy, 6–9 June 2017. pp. 1–6
11. I. Bolvashenkov, J. Kammerman, T. Lahlou et al., Comparison and choice of a fault tolerant inverter topology for the traction drive of an electrical helicopter. in *Proc. of IEEE 4th Int. Conference on Electrical Systems for Aircraft, Railway, Ship Propulsion, and Road Vehicles & International Transportation Electrification Conference (ESARS-ITEC)*, Toulouse, France, 2–4 November 2016, p. 6
12. I. Bolvashenkov, J. Kammermann, S. Willerich et al., Comparative study for the optimal choice of electric traction motors for a helicopter drive train. in *Proc. of 10th Int. Conference on Sustainable Development of Energy, Water and Environment Systems (SDEWES)*, Dubrovnik, Croatia, 27 Sept.–3 Oct. 2015, pp 1–15
13. G.V. Brown GV, Weights and efficiencies of electric components of a turboelectric aircraft propulsion system, in *Proc. of 49th AIAA Aerospace Sciences Meeting*, Orlando, Florida, 4–7 January 2011, pp 1–18
14. T. Brunner, M. Kampitsch, O. Kircher, Cryo-compressed hydrogen storage, in *Fuel cells: data, facts, and figures*. ed. by D. Stolten, R.C. Samsun, N. Garland (Wiley, Veinheim, Germany, 2016), pp. 162–177
15. R. Del Rosario, A future with hybrid electric propulsion systems: a NASA perspective. In *NASA Turbine Engine Technology Symposium*. Strategic Visions Workshop, Dayton, Ohio, 11 September 2014, pp. 1–21
16. R. Del Rosario, Next generation aircraft electrical power systems & hybrid/all electric aircraft. NASA Aerospace Electrical Systems Expo, Long Beach, CA **2015**, 1–12 (2015)
17. S. Eaves, J. Eaves, A cost comparison of fuel-cell and battery electric vehicles. J. Power Sources **130**, 208–212 (2004)
18. L. Eudy, M. Post, *American Fuel Cell Bus Report Evaluation: Second Report*. National Renewable Energy Laboratory, Denver, Co, September 2015, p. 45
19. M. Frank, P. van Hasselt, P. Kummeth et al., High-temperature superconducting rotating machines for ship applications. IEEE Trans. Appl. Supercond. **16**(2), 1465–1468 (2006)
20. I. Frenkel, I. Bolvashenkov, H.G. Herzog et al., Performance availability assessment of combined multi power source traction drive considering real operational conditions. Transp Telecommun **17**(3), 179–191 (2016)
21. I. Frenkel, I. Bolvashenkov, H.G. Herzog et al., Operational sustainability assessment of multi-power source traction drive, in *Mathematics Applied to Engineering*. ed. by M. Ram, J.P. Davim (Elsevier, London, UK, 2017), pp. 191–203
22. L. Jörissen, AutoStack-CORE – Development and tests of the 95 kW fuel cell stack. https://www.zsw-bw.de/en/projects/h2-und-brennstoffzellen/autostack-core-development-and-test-of-a-95-kw-fuel-cell-stack.html. Accessed 08 February 2017
23. Z. Jun, X. Feng, C. Wei et al., The study and test for 1MW high temperature superconducting motor, CSC & ESAS European Superconductivity News Forum (ESNF). Iss. **22**(2012), 1–4 (2012)

24. P. Kurzweil, *Brennsoffzellentechnik: Grundlagen, Komponenten, Systeme, Anwendungen, (Fuel Cell Technology: Basics, Components, Systems, Applications)* (Springer, Wiesbaden (in German), 2013)

25. C.A. Luongo, P.J. Masson, T. Nam et al., Next generation more-electric aircraft: a potential application for HTS superconductors. IEEE Trans. Appl. Superconduct. 19(3), Part 2: 1055–1069

26. N. Madavan, J. Heidmann, C. Bowman, A NASA perspective on electric propulsion technologies for commercial aviation, in *Workshop on Technology Roadmap for Large Electric Machines*, University of Illinois, Urbana-Champaign, 5–6 April 2016, pp. 1–32

27. P. Malkin, M. Pagonis, Superconducting electric power systems for hybrid electric aircraft. Aircr. Eng. Aerosp. Technol. 86(6), 515–518 (2014)

28. P.J. Masson, G.V. Brown, D.S. Soban et al., HTS machines as enabling technology for all-electric airborne vehicles. Supercond. Sci. Technol. 20(8), 748–756 (2007)

29. S. Matlok, Bidirectional full SiC 200 kW DC-DC converter for electric, hybrid and fuel cell vehicles. Fraunhofer Institut for IISM, p. 2

30. G.J. Offe, D. Howey, M. Contestabile et al., Comparative analysis of battery electric, hydrogen fuel cell and hybrid vehicles in a future sustainable road transport system. Energy Policy 38(1), 24–29 (2010)

31. Rick H (2013) Gasturbinen und Flugantriebe: Grundlagen, Betriebsverhalten und Simulation. Springer, Berlin (In German)

32. D. Rotenberg, A. Vahidi, I. Kolmanovsky, Ultracapacitor assisted powertrains: modeling, control, sizing, and the impact on fuel economy. IEEE Trans. Control Syst. Technol. 19(3), 576–589 (2011)

33. R. Schiferl, A. Flory, W.C. Livoti et al., High-temperature superconducting synchronous motors: economic issues for industrial applications. IEEE Trans. Ind. Appl. 44(5), 1376–1384 (2008)

34. R.M. Sullivan, J.L Palko, R.T. Tornabene et al., Engineering analysis studies for preliminary design of light weight cryogenic hydrogen tanks in UAV applications. NASA/TP—2006–214094, p. 27

35. C.E. Thomas, Fuel cell and battery electric vehicles compared. Int. J. Hydrogen Energy 34, 6005–6020 (2009)

36. P. Thounthonga, S. Raël, B. Davat, Energy management of fuel cell-battery-supercapacitor hybrid power source for vehicle applications. Int. J. Power Sources 193, 376–385 (2009)

37. L.S. Yanovsky, V.V. Krimov, A.G Finogeev, Alternative power units for aircraft traction drives on the basis of fuel cells, in *Proc. of the Moscow Aviation Institute (Trudi MAI)*, Iss. 56, 2012, pp. 1–6 (In Russian)

38. T. Yoshida, K. Kojima, Toyota MIRAI fuel cell vehicle and progress toward a future hydrogen society. Electrochem. Soc. Interface 24(2), 45–49 (2015)

Chapter 4
Electric Propulsion Systems of Interorbital Rockets for Flights Toward Planets of the Solar System

4.1 Introduction

The conquest of the interplanetary space dreamed of by the pioneers of astronautics has begun in the twentieth century. The successful use of rockets with chemical rocket engines has led to the mastering of near-Earth space and to expeditions to the Moon. The main achievement of the twenty-first century is the creation of an international orbital space station (ISS), which can become a port from which moorages the spacecrafts can go toward the planets of the Solar System.

The next step in the development of astronautics should be the implementation of scientific expeditions toward the planets of the Solar System and their satellites.

However, chemical rocket engines have a specific energy impulse, which is an order of magnitude lower than that of an electric rocket engine, and they are not capable of carrying out interorbital rocket flights. Electric rocket engines are designed for this purpose. The greatest tractive force for the implementation of jet movement is created by a magnetoplasma electric rocket engine (MPE).

When this engine works, a discharge burns between the electrodes, and the interaction of the current with the magnetic field creates the force of thrust. The engine of this type was first developed at Princeton University in the United States and tested at NASA's Jet Propulsion Laboratory [4].

A detailed description of the MPE is provided in the work of Princeton University professor Choueiri [2]. Unfortunately, the efficiency of MPE engine of a known design is only 60%.

In 2006, the author of this chapter has proposed a new design of the MPE engine, which has an external superconductor winding, creating a tangential field in the working chamber. To do this, the current in the superconductor winding is directed along the engine axis. This has increased the efficiency of the engine from 60 to 92% [8]. In 2008, the design of a reusable spacecraft for the mission to Mars was developed. Interorbital flight is carried out with the help of a superconducting electric rocket engine, and landing and taking off from the surface of Mars are carried out with the help of a chemical rocket engine.

© The Author(s), under exclusive license to Springer Nature Switzerland AG 2022   63
I. Bolvashenkov et al., *Vehicle Electrification*, SpringerBriefs in Applied
Sciences and Technology, https://doi.org/10.1007/978-3-030-81740-4_4

In 2008, a demonstration model of the spacecraft for the mission toward Mars was made.

In 2011, the design of a space train for flights to the planets of the Solar System was developed. The space train consists of a locomotive with electric rocket engines and an onboard power plant, and the working substance hydrogen in liquid state is in a tank container.

A new design of onboard power plant consisting of a gas nuclear reactor, a magnetohydrodynamic (MHD) generator, and a turbogenerator has been developed. For landing and taking off from the surface of the planets and their satellites, the design of the takeoff–landing capsule (TLC), inside which is the cabin for astronauts, was developed. The locomotive, the tank container, and a takeoff–landing capsule are being connected in the orbits of the planets of the Solar System by docking, forming a space train.

In 2014–2016, the task of carrying out expeditions toward all the planets of the Solar System was solved. For this purpose, a technique was proposed, and a new design of the interorbital rocket train was developed. The rocket train is supplemented by a refueler with a tank of working substance, which during the preliminary flight is sent into the orbit of the planets of Mars, Venus, and Mercury and is used for the return flight of the expedition.

For expeditions to the gas planets of the Solar System: Jupiter, Saturn, Uranus, and Neptune, the design of an autonomous refueler of working substance was developed, which during preliminary flight is sent to the satellites of these planets: Europe, Titan, Oberon, and Triton, the surface of which is covered with water ice. The autonomous refueler melts ice, and out of liquid, water by means of electrolysis produces hydrogen, which after liquefaction is refueled into the tank containers of the space train.

In 2016, a study of the superconductive magnetoplasm electric rocket engine was conducted [8]. A mathematical model of electromechanical processes taking place in the working chamber was drawn up. The study confirmed the effectiveness of the technical decisions made. A new electric motor design has also been proposed, which, by moving and cooling of the cathode, allows to increase for several times the lifespan of the engine.

For the research, a mathematical model of the electric propulsion system was developed, in which superconducting electric machines (electric generators and electric motors) work jointly.

The fulfilled projects of expeditions toward the planets of the Solar System have shown that the most expensive part of the space train is the onboard power plant consisting of a nuclear reactor, an MHD generator, and a turbogenerator. A space train project, which electric rocket engines are powered from solar panels, has been also considered. In this design, the cost of creating onboard supply units is significantly reduced.

As for the solar panels themselves, one should notice that this field has undergone impressive changes in recent years, and factually, we are witnessing a technological revolution in energetics that takes place based on solar panels. The efficiency of solar panels made from silicon has substantially improved.

Results of modern researches given in [8] show that the efficiency of the solar cell has reached 23%, and the solar panel's specific weight reaches 2.5 kg/m^2.

Significant progress has been made in the development of gallium arsenide-based solar cells. Today, the world market includes photovoltaic converters Ga-As, manufactured by Electrolab (USA), Soitec (Germany), Sharp (Japan), and others based on multilayered structures. The technology of a three-cascade solar cell based on gallium arsenide with a silicon substrate has been developed. The efficiency of solar panels produced has reached 44%.

In 2017, a study was carried out, which showed that the specifications of the industry-made solar panels allow to provide the power needed for the electric movement of interorbital rockets. A new solar locomotive design has been developed that combines a tank with a working substance and a solar battery in one rocket. Calculations of the processes taking place during the movement of the space train, made on a mathematical model, showed that the simplification of the design and the reduction in the cost of the onboard supply unit makes the flight to Mars a reality of the near future.

The objectives of this study are as follows:

- based on consideration of projects of rockets carrying out expeditions to the planets of the Solar System, to identify the main design solutions allowing to build space trains capable of moving in the Solar System at maximum speed.
- to conduct research and find the optimal solution to the generating part of the electric propulsion system, both based on solar panels and on the nuclear power unit.
- using the high specifications of the SERPS, to develop a system of electric propulsion of space trains of the future, which will allow to conduct scientific expeditions to all planets of the Solar System.
- to lay the foundations of the interplanetary communication system of the future.

4.2 Superconducting Electric Rocket Propulsion System (SERPS)

According to the principle of action, the SERPS belongs to the class of magnetohydrodynamic DC machines and is an electromechanical energy converter, the action of which is associated with the movement of gas in plasma state with the help of a magnetic field.

To do this, the gas feeds into the space between the positive and negative electrode, which serves as a working chamber.

The electrodes are being connected to an onboard power source with a voltage "U", and in the engine's working chamber, an electric field is created.

Under the influence of the electric field, the ionization of the gas takes place (i.e., removal of electrons from gas molecules). As a result, the molecules turn into positive ions.

After the formation of electrons and positive ions, the gas, which is the working substance of the electric motor, is in state of plasma.

The value of a direct current, flows through the plasma, may be determined using the following ratio:

$$I = \frac{U}{R},\qquad(4.1)$$

where R—electrical resistance of the plasma,

U—voltage of onboard power source.

In 2006, a new design solution was found in the process of development of the engine: Outside of working chamber, a superconducting excitation winding was installed, creating permanent magnetic field inside the working chamber. This excitation winding is reeled in such a way that the current in it runs parallel to the axis of the electric motor and creates the magnetic field directed perpendicular to the current between the electrodes.

Magnetic field tension is determined by the law of Biot–Savart–Laplace

$$B_\tau = I_B k_i \qquad(4.2)$$

where: I_B—current of the excitement in the superconducting winding,

k_i—a geometric coefficient, which is found after solving the task of magnetic field distribution in the work chamber.

In accordance with Ampère's law, after switching on the superconducting winding an electromagnetic force dF acts on the elementary plasma volume dV and is equal to:

$$\mathrm{dF} = \left[\overline{J} \times \overline{\mathrm{B}}_\tau\right]\mathrm{dV} \qquad(4.3)$$

where $\overline{J}$—plasma current density.

The full magnitude of this force, which determines the thrust of the electric motor, can be represented in the form of

$$F = \int_V \left[\overline{J} \times \overline{B_\tau}\right]\mathrm{dV} = I B_\tau l, \qquad(4.4)$$

where l is the distance between anode and cathode, and I is the value of current.

Under the action of the arising force F, plasma begins to move along the horizontal axis of the working chamber at the speed of expiration V, at which there is balance of voltages:

$$U - E = I R \qquad(4.5)$$

where E is electromotive force (EMF), arising between electrodes and determined by Faraday law:

$$E = B_\tau l \cdot V \tag{4.6}$$

and IR—voltage drop when the current passes through the plasma.

Under the action of the F force, plasma is ejected from the working chamber, creating the thrust of the electric rocket engine.

The electric motor is mounted on a rocket that is in space, and according to Newton's third law, it experiences the impact of this force F (with a reverse sign in the direction of movement).

In accordance with Ohm law, the electric resistance of the engine is

$$R = {}^l/_{\sigma S} \tag{4.7}$$

where S—cross section of the plasma jet between the anode and the cathode and is calculated as follows:

$$S = \pi (d_k + l) L_k, \tag{4.8}$$

where d_k—cathode diameter; L_k – cathodelength; σ—plasma conductivity.

It should be recalled that theoretical and experimental studies of plasma, as a conductor of electric current, were conducted by scientists of different countries 50 years ago in the creation of magnetohydrodynamic (MHD) generators. In particular, the mechanism of plasma electrical conductivity in the crossed magnetic and electric fields was considered in [9]. It has also been found that under the influence of the magnetic field a decrease in the electrical conductivity of the plasma takes place. It has been shown in the paper that significant increase in plasma electrical conductivity can be obtained through additives from easily ionized substances, such as cesium. In 2011, a study was conducted on the electrical conductivity of molecular hydrogen plasma with alkaline metal additives. The advantages of molecular hydrogen, allowing it to be used as a working substance in MHD generators, have been revealed [3, 9]. Based on the results of the studies [5], it is shown in Fig. 4.1 how the electrical conductivity coefficient of nitrogen plasma 1 and hydrogen plasma 3 changes when the temperature and the magnitude of the applied magnetic field change.

Hydrogen and nitrogen, used as working substance of the electric motor, when stored in a liquid state, become the medium allowing the work of superconducting excitement windings.

The magnitude of the magnetic induction of the superimposed magnetic field is 1.5 T (curve 2—nitrogen, curve 4—hydrogen).

Fig. 4.1 Plasma electrical conductivity coefficient

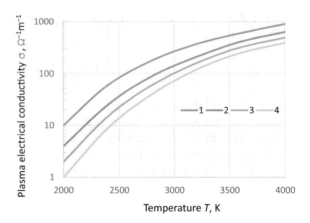

During the design, calculations were made of the parameters of electric motors for the flight to the orbits of the planets of the Solar System, which showed that the optimal induction of the transverse magnetic field B_τ should be 1.2–1.5 T.

The optimization criterion was to obtain minimum mass of electric rocket. When calculating electric motors during design, the optimization program also determined their efficiency.

In so doing, the energy losses in current flow through the electroconductive plasma were determined by the law of Joule–Lenz:

$$\Delta P = I^2 R \tag{4.9}$$

The efficiency of the electric motor is

$$\eta = 1 - \Delta P / P \tag{4.10}$$

where $P = UI$ is the power that is supplied to the engine from the onboard power source.

4.3   Design of SERPS

The longitudinal section of the engine is presented in Fig. 4.2.

In the working chamber along the engine axis, a cathode 1 and an anode 2 (outside), which have a cylindrical shape, are located. The cathode 1 is fixed on the anode 2 with the help of a cylindrical sleeve 3 and an insulator 4, which has the shape of a cylinder.

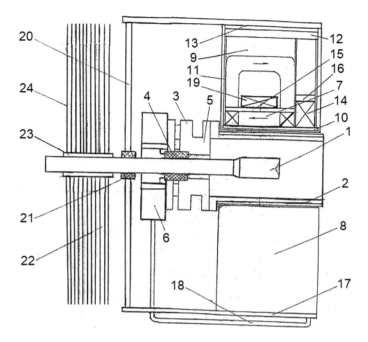

Fig. 4.2 Longitudinal section of SERPS

The working substance of the electric motor can be both nitrogen and hydrogen, which in a gaseous state are fed along channel 5 in the sleeve 3 into the interelectrode space. The gas is delivered by pipeline 18 to the working substance conditioning chamber 6, which is ring-shaped. Inside chamber 6 is a dispenser and an electromagnetic valve. The anode 2 is installed and fixed in the cylindrical cavity of cryostat 8, which is made of carbon and where the magnetic system of the electric motor is located.

The superconducting magnetic system consists of three windings of different purpose.

The first winding 7 consists of 4 coils, created a tangentially directed magnetic field in the engine's working chamber.

The second winding 14 has cylindrical shape and is designed to compress the working substance, which in state of plasma is ejected from the working chamber. The third winding 19 has cylindrical shape and is designed to stabilize the electrical discharge between electrodes 1 and 2 by evenly rotating the electric arc.

Figure 4.2 shows elements of intense cooling of the cathode 1. For this purpose, on the opposite side of the working chamber on the surface of the cathode with the help of cylindrical pad 23, a disk radiator 24 is installed, which gives heat to outer space.

Reducing the surface temperature of the cathode reduces the size of the cathode mass carryover and increases the lifespan of the electric motor.

Fig. 4.3 Longitudinal section of the magnetic field distribution scheme of SERPS

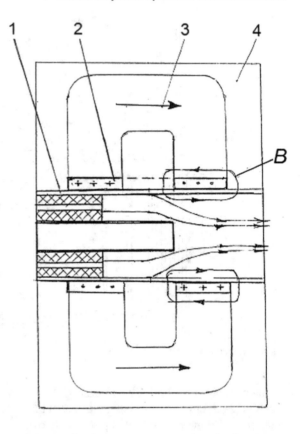

The functioning of the superconducting magnetic system of the developed SERPS is explained by the magnetic field distribution scheme shown in Fig. 4.3 in longitudinal section and in Fig. 4.4 in cross section.

In these figures, numbers are as follows: 1—internal cryostat of cylinder, 2—half-coil of the superconducting excitation winding, 3—half-coil of the excitation winding with reverse current, and 4—radial rib of cryostat.

As it can be seen in Figs. 4.3 and 4.4, four half-coils, in which the current is directed parallel to the longitudinal axis of the engine, create inside the working chamber the magnetic field directed perpendicular to the current between the electrodes, which has radial direction.

The calculated parameters and sizes of the electric motor for the movement of the rocket from Earth's orbit to the orbit of Mars are shown in Table 4.1.

Fig. 4.4 Radial cross section of the magnetic field distribution scheme of SERPS

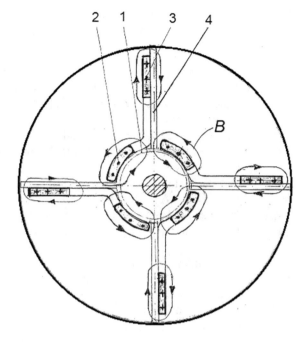

Table 4.1 Parameters of SERPS

Parameters and sizes of the electric motor	
Traction force	60 N
Power	600 kW
Current	600 A
Voltage	1000 V
Efficiency	94.5%
Specific pulse	6000 s
Expenditure of the working substance	1 g/s
Exhaust velocity of the working substance	10 km/s
Magnetic induction	1.5 T
Anode diameter	170 mm
Cathode diameter	40 mm
Anode length	100 mm
Cathode length	60 mm
Diameter of external cylinder	650 mm
Length of external cylinder	450 mm

4.4   Electric Rocket for Movement of a Space Train from Earth's Orbit to Mars' Orbit

Calculation and design studies conducted on the mathematical model have showed that expeditions to the planets Mars, Venus, and Mercury should be carried out with the help of space trains, which are collected in Earth' orbit from electric rockets and takeoff–landing modules.

The number of electric rockets and modules in each train depends on its purpose. Each component of the space rocket train is put into circumterrestrial orbit by the "Delta IV Heavy" rocket carrier.

The interorbital flight of the space train is carried out by a rocket, which is put in movement by a SERPS. A sliding solar battery based on gallium arsenide is installed on the electric rocket to power electric motors. Working substance—nitrogen for operation of electric motors—is placed in a cryogenic tank located along horizontal axis.

The takeoff–landing modules, which house the expedition's crew cabin, are equipped with a hydrogen–oxygen chemical rocket engine, which allows the module to move around the Earth and around Mars.

Studies have shown that an expedition to Mars is appropriate to conduct in two stages. In the first stage, a working substance, nitrogen in liquid state for the operation of electric rocket engines, is delivered into orbit around Mars, providing the movement of the space train with the crew of the expedition from the orbit of Mars to the Earth's orbit. The scheme of interorbital space train no.1 for the first stage is shown in Fig. 4.5.

The flight is carried out with the help of an electric rocket ER-7, along the axis of which the tank 4 with liquid nitrogen is being docked. The rocket is moving along interorbital trajectory using four SERPSs 6, which are powered by solar cell 1. After assembling space train no.1 in Earth's orbit in automatic mode, with the help of docking unit 5, its movement on the calculated trajectory shown in Fig. 4.6 (curve 1) begins.

The movement trajectory of the space train is built with the help of an astrodynamic program, which continuously determines the forces of interaction of the space train with the Sun, Earth, its satellite Moon, and the planets Mercury, Venus, and Mars as they move together along their orbits [7].

In so doing, the picture of the gravitational field of the Solar System in the zone of movement of the space train, which is the body of variable mass, being under the influence of this field, is determined. The calculation program also determines the change in the thrust of rocket engines depending on the distance to the Sun. This factor is taken into account at each point of the trajectory using the method of sequential approximations. It should not be forgotten that the intensity of solar radiation is determined by the law of reverse squares and is 1370 W/m^2 in the Earth's orbit. In Mars' orbit, this value is reduced by 2.3 times.

Train no.1 (Fig. 4.5) starts from point 4 in Earth's orbit. After accelerating up to speed of 35 km/s during 6 days for 126 days, the flight is continued with the engines

Fig. 4.5 Space train no.1 for the first stage an expedition to Mars

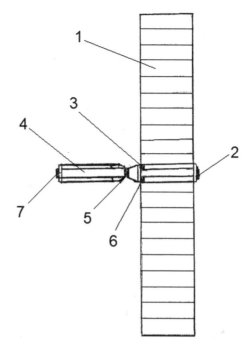

Fig. 4.6 Movement trajectory of the space train to Mars

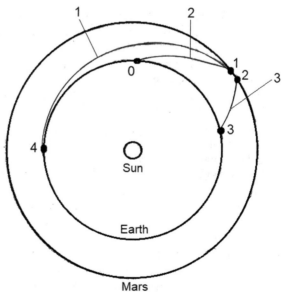

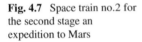

Fig. 4.7 Space train no.2 for
the second stage an
expedition to Mars

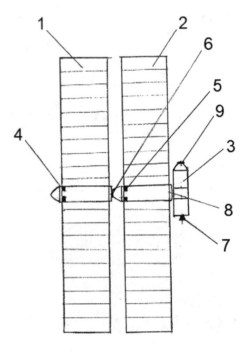

6 turned off. Then, the electric rocket engines are turned on in braking mode for 6 days. The speed of the train decreases to 5 km/s, and it enters orbit around Mars in point 1. At the same time, the mass of the working substance (nitrogen), which is in tank 3, is reduced from 14 to 11.5 tons. In the future, this reserve of the working substance will be used for the return flight of the expedition to the Earth's orbit.

In the second stage, space train no.2 (Fig. 4.7) is sent from Earth's orbit to Mars' orbit.

It is assembled in Earth's orbit by docking two electric rockets ER-7, 1 and 2. Electric rocket ER-7 is delivered in assembled form being in the head part of the Delta IV Heavy rocket carrier.

After unfolding of solar panels, docking of rockets 1 and 2 is made. Description of the ER-7 design is given below.

It is in the 26-ton takeoff–landing capsule (TLC) where the expedition team, consisting of three astronauts and 18 tons of liquid hydrogen and oxygen needed for the operation of the chemical rocket engine 7 when taking off from the surface of Mars, is located.

When moving from Earth's orbit to Mars' orbit, nitrogen in the tanks of rockets 1 and 2 is fully consumed, and nitrogen to conduct the return flight delivers rocket train no.1. Therefore, the launch of space train no.2 should be carried out 90 days after the launch of space train no.1.

The movement of train no.2 begins with the delivery of astronauts to the International Space Station. In so doing, the well-known delivery systems, such as Soyuz or

Space X, are used. Astronauts are waiting for the beginning of the interorbital flight being at the ISS.

The takeoff–landing capsule (TLC) (Fig. 4.7) 3 is launched into Earth's orbit with the "Delta IV Heavy" rocket carrier. After the launch, the TLC goes to the convergence and mooring to the ISS in automatic mode. With the help of docking unit 9 in the nasal part of the TLC, the connection with a moorage of the ISS takes place. After boarding of astronauts, the takeoff–landing capsule with the help of a chemical rocket engine 7 is approaching the electric rocket 2. Using the docking unit 8, which is located on the transverse axis of the capsule, the docking of rocket 2 and capsule 3 is carried out.The sentence , With the help of docking unit 9 in the nasal part of the TLC connection with a moorage of the ISS takes

The system creation of artificial gravity that occurs in the cabin of the TLC when rotating around the transverse axis with the help of superconductor bearing 8 is being switched on.

The astronaut-machinist switches on electric rocket engines 4, the space train begins to increase the speed of movement, and after reaching the second space, speed goes into the calculated trajectory of the flight (curve 2 in Fig. 4.6).

Figure 4.8 shows the process of changing the speed (V), mass (M) of the train, and the power (P) of the onboard energy source over time.

As it can be seen from Fig. 4.8, the acceleration of the space train no.2 to the speed of 100 km/s occurs within 22 days. At this time, the electric rocket engines operate with maximum power, which is continuously reduced. Then, the electric motors turn off, and the train moves by inertia for three days. Then, the rockets make 180-degree U-turn. After turning on the electric motors, the space train reduces the speed of movement during 15 days from 100 to 3.6 km/s.

The space train becomes a satellite of the planet Mars. In the orbit of Mars, the meeting of train no.1 and train no.2, their rapprochement, and docking occur. The orbital flight took 40 days. The next stage of the movement is the landing of the takeoff–landing capsule on the surface of Mars. But first you need to refuel cryogenic tanks of rockets 1 and 2 by the working substance of electric rocket engines 4 and 5 (nitrogen), which is in liquid state in the tank refueler 4 (Fig. 4.5).

The astronaut, who is in the cabin of the takeoff–landing capsule, gives the command to turn on cryogenic pumps located in the cryogenic systems of rocket 1 and rocket 2. The liquid nitrogen is pumped over from the tank of the refueler 4 (Fig. 4.5) into the tanks of rockets 1 and 2 (Fig. 4.7).

Landing of takeoff–landing capsule on the surface of Mars begins with the fact that the astronaut-machinist reduces the speed of the TLC. TLC leaves space trains no.1 and no.2, which remain in orbit around Mars.

At a speed of 3.5 km/s, the TLC moves toward the surface of Mars with an acceleration of 0,38 g. When entering the atmosphere of Mars at a distance of 60 km from its surface, the parachute system triggers. Takeoff–landing capsule 3 (Fig. 4.7) is approaching the surface of Mars. Using the chemical rocket engine **7**, the astronaut makes smooth landing at the point that has been determined by the expedition plan.

After landing, the astronauts go into the surface of Mars. Using the latest instruments, astronauts conduct a complex of experiments for 8 days.

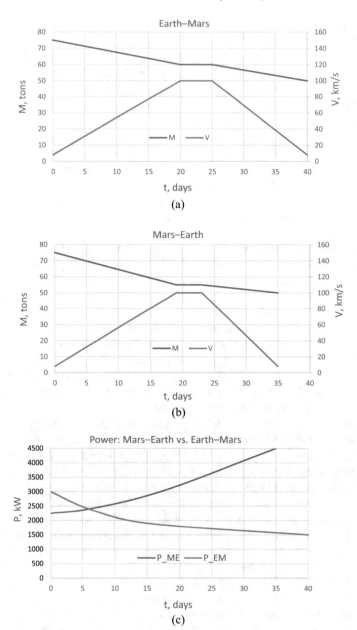

Fig. 4.8 Speed (**a**), mass (**b**) of the train, and the power (**c**) of the energy source during flight

When the planned works are completed, the astronauts take places in the cabin of the takeoff–landing capsule. The chemical rocket engine 7 is switched on. The takeoff–landing capsule breaks away from the surface of Mars and in 220 s enters orbit around Mars, on which the space train no.3 (Fig. 4.9) is moving.

The space train no.3 is designed to provide for flight of the expedition from Mars' orbit to Earth's orbit. The space train no.3 is assembled from three ER-7 rocket modules, which are in Mars' orbit, by maneuvering with the help electric rocket engines 4 and 10.

After docking the rocket train with the takeoff–landing capsule 3 with the help of docking unit 8, the assembly ends. As a result, we got a rocket that has three solar panels of 1, 2, 6, and 12 electric rocket engines of SERPS type 4, 5, 10.

After turning on the electric rocket engines by the astronaut, the rocket train 3 gain up speed, leaves the orbit of Mars and takes the course onto the Earth's orbit. The trajectory of the rocket train no.3 is shown in Fig. 4.6. The movement of the rocket train begins at the moment when Mars is in its orbit around the Sun at point 2. When moving, the space train must pass from point 2 to point 3, where the planet Earth will be at the time of completion of the movement. The results of the calculations of the train parameters in time are shown in Fig. 4.8. During the 18 days of flight,

Fig. 4.9 Space train no.3 for the flight from Mars' orbit to Earth's orbit

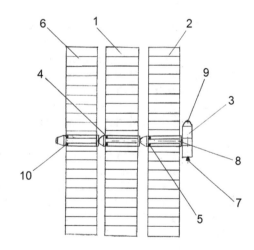

Fig. 4.10 ER-7 rocket

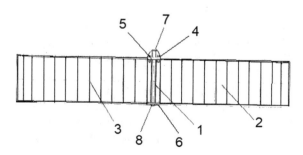

the space train accelerates up to speed of 100 km/s. This is due to the fact that when approaching the Sun, the power of solar panels of three connected rockets 1, 2, and 6 continuously increases and after 20 days becomes equal to 3.0 MW.

After 18 days, the electric rocket engines are switched off, and the space train continues to move by inertia at speed of 100 km/s for 8 days. After the maneuver of the space train, the thrust direction of the electric rocket engines changes by 180 degrees. The braking of the train begins with a reduction in speed to 10 km/s for 8 days. The return flight from Mars' orbit to Earth's orbit takes 34 days. After entering the Earth's orbit, the space train is uncoupled. Electric rockets ER-7 1, 2, and 6 are refilled with working substance—nitrogen. In the future, they are used for re-flying Mars. Takeoff–landing capsule 3 with a crew onboard with the help of chemical rocket engine 7 (Fig. 4.9) is moving into the orbit of the International Space Station (ISS). The capsule steers to an ISS moorage and connects with it with the help of docking unit 9. The crew of the expedition goes from the capsule to the ISS.

The expedition to the surface of Mars, which lasted 82 days, ends.

4.5 Design of the Electric Rocket ER-7

The ER-7 rocket is named after the famous R-7 rocket, which was created under the direction of Sergey Korolev, and has carried out the first flights into space (Fig. 4.10).

Here, 1—case, 2—the right solar battery, 3—the left solar battery. Solar batteries have panels consisting of cells made of gallium arsenide. 4—electric rocket engine SERPS, 5—front end disk, 6—rear end disk, 7—front docking unit, and 8—rear docking unit.

The design of the rocket case and its connection with the solar battery are shown in Fig. 4.11.

The solar battery has a folding structure. It is assembled from separate panels 4, which are made from carbon fiber. Outside in the panel, a layer of photovoltaic converter embedded, which is made of gallium arsenide. The panels are connected with the help of hinges 16, 17. The cross sections of the case design are shown in Figs. 4.12 and 4.13.

Figure 4.12 shows the position when the solar panel is in the folded form. While being put into orbit around the Earth, the structure is held by a cylindrical shell. Installation of solar panels in working position is carried out with the help of sliding rods fixed on longitudinal beams 5, 6.

Figure 4.13 shows the fastening of sliding rods 6 and 7 inserted into each other. The cylindrical case of the rocket 1 (Fig. 4.11), made of aluminum alloy, serves as the basis of the design. On both sides of the case 1, end disks 2 are set. The tank 3, which stores liquid nitrogen, is made of carbon fiber and has on-screen vacuum insulation. In the end tank 3, there is a cryogenic pump 8 for the pumping liquid nitrogen. In the end disk 2 (Fig. 4.11), electric rocket engines 6 are installed. In Fig. 4.13, one can see that the end disk has four holes located at angle of 90°. In the bow of the

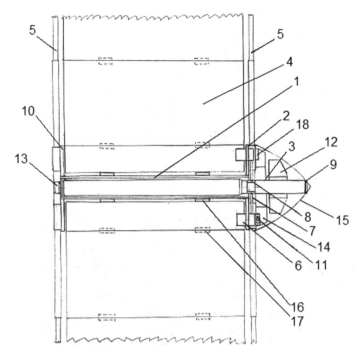

Fig. 4.11 Design of the ER-7 rocket and its connection with the solar battery

Fig. 4.12 Cross section of the case design: solar panel is in the folded form

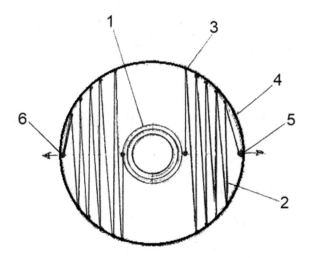

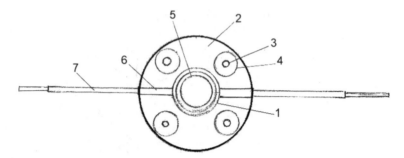

Fig. 4.13 Cross section of the case design: fastening of sliding rods 6 and 7 inserted into each other

rocket, the outer cone 15 is installed, in the inner cavity of which blocks of automatic control system 12 are located.

The geometric dimensions of the ER-7 rocket are: case length 25 m, case diameter 1.1 m, the end disk diameter 6 m. The length of the solar battery wing is 60 m. The width of the solar battery wing is 20 m. The space train has an autonomous control system. The control system in the form of blocks is placed in container 12. Changing the direction of the train—pitching and hunting—is carried out by changing the traction of electric motors 6. Changes in the traction force are made by changing the electric motor's current with the help of a power transistor. The rocket rotates with the help of an electric rocket engine 11 of low traction.

This electric motor, which creates a tangential thrust of 10 N, has a power of 80 kW. The electric rocket engine 11 is the executive element of the system 18 constantly tracking of the position of the Sun, which is installed on the end disk 2. System 18 determines the angle of rotation of the rocket relative to the longitudinal axis, at which the solar panel produces maximum power.

4.6 Design of the Takeoff–Landing Capsule (TLC)

The capsule is designed to move three astronauts from Earth's orbit to Mars' orbit, for landing on the surface of Mars, for taking off into its orbit and to flying from Mars' orbit to Earth's orbit. The design of the capsule is shown in Fig. 4.14.

Here,

1—chemical hydrogen–oxygen rocket engine with a thrust force of 12 tons and a specific pulse of 400 s,

2—bottom of the capsule's rocket case,

3—fuel tank with liquid oxygen,

4—fuel tank with liquid hydrogen,

5—chemical rocket engine for maneuvering,

6—tank with liquid hydrogen,

7—cabin of expedition crew,

Fig. 4.14 Design of the
takeoff–landing capsule

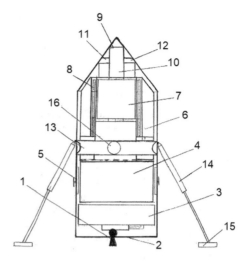

8—superconducting solenoid, designed to create a permanent magnetic field in
the cabin of expedition crew,

9—docking unit on the longitudinal axis,

10—tunnel and gateway to exit the rocket,

11—parachute system for landing on the surface of Mars,

12—container for measuring devices,

13—load-bearing ring of the landing tripod,

14—damper of landing support,

15—shoe of landing support,

16—docking unit on the transverse axis.

TLC mass on the Earth—26 t, TLC length—18 m, TLC external diameter—5.8 m.

4.7 Electric Rocket to Move Space Train from the Earth's Orbit to Orbit of Jupiter

Returning to the problem of expeditions to the distant planets of the Solar System
Jupiter, Saturn, Uranus, and Neptune, one should recognize that at the moment their
implementation remains a dream. The reason for this is the lack of sufficient progress
in development of the rocket electrical propulsion systems.

This chapter presents a new constructive and technological solution. It should be
noted that expeditions to distant planets will require the transfer of large masses over
long distances with maximum speed. It is almost impossible to do this using solar
energy. In the orbit of the planet Jupiter, the intensity of solar radiation due to the
large distance from the Sun is reduced by 27 times as compared to the Earth.

For flights to the planet Jupiter, a system of electric propulsion of the rocket train,
which is collected in the orbit of the Earth from individual elements by sequential

docking, is developed. The separate elements: the space locomotive, tank containers with working substance, and the takeoff–landing capsule with a cabin for astronauts are placed into the Earth's orbit with the help of carrier rocket «Arian-5». The scheme of an interorbital rocket train for flight to Jupiter is shown in Fig. 4.15.

The movement of the carrier rocket is provided by means of the locomotive 1, which has electric rocket engines 5. For feeding of the electric rocket engines 5, the power supply—the nuclear reactor with MHD generator and turbogenerator, is installed aboard of the locomotive. The working substance for electric rocket engines 5 (liquid hydrogen) is stored in the tank containers 2, 3, which are docked the locomotive 1 along the horizontal axis.

The capsule 4 is the useful load of the carrier rocket and is intended for landing at the solid surface of the Jupiter satellite Europa after entering its orbit, as well for taking off the surface of the satellite and entering into its orbit after the end of the stay. These operations are performed by means of the chemical rocket engine 9. Inside of the capsule 4, the crew cabin is located. In order to provide at the process of the expedition the artificial gravitation in the cabin, the capsule rotates relative to its transverse axis.

In the search process of the new concept for realization the flight to Jupiter, the attention was paid to the geological structure of the Europe. According to the data of the newest research, the surface of the satellite Europe is covered with the layer of the water ice. If to convert this ice to liquid water and to obtain gaseous hydrogen from the liquid water, it is quite possible to convert the gaseous hydrogen to the liquid state and to deliver it from the satellite Europe surface into its orbit. In this case, there is no need for the preliminary delivery of the working substance from the Earth's orbit. A stand-alone apparatus, which is able to perform all the necessary operations for filling of the carrier rocket with the working substance, shall be delivered into the Europe orbit.

The preliminary scheme of the carrier rocket with such a refueler is shown in Fig. 4.16.

For the purpose of the refueler delivery into the Europe orbit, a space train is formed at the Earth's orbit. The space train consists of the locomotive 1, having the electric rocket engine 2, two tank containers 3 and 4 with the working substance, connected by means of the docking unit 5, and the stand-alone space refueler 6.

Fig. 4.15 Interorbital rocket train for flight to Jupiter

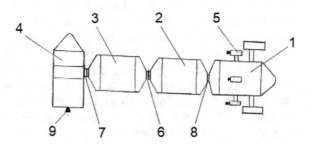

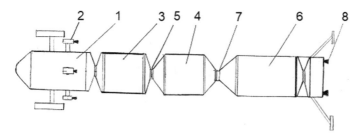

Fig. 4.16 Rocket carrier with refueler

The refueler 6 is launched out of the Earth surface and is put into its orbit. Then, by means of the docking unit 7, the space refueler is connected with the tank container 4.

By means of the locomotive 1, the space train reaches the Europa orbit, and during this flight the working substance, which is found in the tank containers 3 and 4, is consumed completely.

On the Europe orbit, the separation of the space train takes place. The locomotive 1 with the empty tank containers remains on the Europe orbit, while the refueler 6 with the help of the chemical rocket engines 8, working in the braking mode, leaves the Europe orbit and carries out the landing on its surface.

With the help of the equipment for liquid hydrogen production out of the water ice, which is installed in the refueler 6, the working substance, which was produced by the refueler equipment itself, is pumped in the hydrogen tank of the refueler.

With the help of the chemical rocket engines, the refueler starts from the Europe surface and orbits it. After approaching the space locomotive 1, the refueler docks to the tank container 4, making use the docking unit 7. Then, the pump for pumping over the working substance is switched on, and the liquid hydrogen, which is found in the refueler, pours into the tank container 4.

After emptying of the refueler tank, the refueler is detached off the tank container 4. Then, the refueling cycle is repeated: The refueler leaves the Europe orbit and carries out the landing on the Europe surface, produces the working substance out of ice and again orbits the Europa, and docks to the tank container 4.

After fueling the second tank container, the refueler 6, with the help of the docking unit 7, detaches off the tank container. The rocket engines 8 are switched on, and the refueler carries out the landing on the Europa surface.

Thereby, the permanent readiness is maintained to fill again the working substance into the carrier rocket and falls away the necessity in carrying out the preliminary flights under fulfillment of the repeated expeditions. The final stage of the preliminary flight is carried out in accordance with the scheme, which is shown in Fig. 4.17.

The space locomotive with the help of the electric rocket engine 2, using the working substance in the tank containers 3 and 4, is going to Earth. After orbiting the Earth, the locomotive 1 is ready to form the carrier rocket, which is shown in Fig. 4.15, and must carry out the main flight to the Europe—the satellite of Jupiter.

Fig. 4.17 Space locomotive
for the final stage of the flight

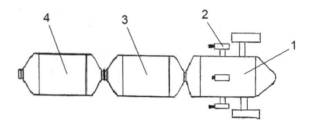

Now, for the flight, it is necessary to form the space train. To this purpose, three launchings are carried out with the help of the carrier rocket "Arian-5". In the first launching into the Earth's orbit, the tank container 2, filled with the working substance, is put. The train formation begins using the electric rocket engines 5. The space locomotive approaches and docks with the tank container 2. Then, in the second launching into the Earth's orbit, the tank container 3 is put and via the docking unit 6 docks the tank container 2.

With the help of the last launching, the takeoff–landing capsule 4 is put into the Earth's orbit. The takeoff–landing capsule 4 is put into the Earth's orbit (without astronauts) by means of the carrier rocket "Arian-5" and with the help of chemical rocket engine 9 begins to approach to the International Space Station (ISS). Then, in an automatic mode, the taxing and docking to one of the ISS docking units are carried out. After the docking, the crew of the expedition, which was delivered beforehand at the ISS, moves into the cabin, located inside the capsule 4. By means of the chemical rocket engine 9, the capsule 4 "pushes off" the ISS and goes into orbit of the space train being formed.

After the docking of the capsule 4 with the tank container 3 (by means of the docking unit 7), the space train, consisting of four elements connected as it is shown in Fig. 4.15, is ready for the flight to Europa. After switching on of the electric rocket engines, the space train speeds up and at attainment of the second space velocity comes into the calculated flight trajectory.

With the help of computerized calculation programs, a search of the optimal calculated trajectory of the flight to Jupiter was carried out. This trajectory has proved to be similar to the trajectory along which the spacecraft "Cassini" had successfully flown in 2000.

The arrangement of the planetary orbits and the calculated flight trajectory are shown in Fig. 4.18, where M—Mars' orbit, E—Earth's orbit, J—Jupiter orbit, and S—the Sun.

At the movement of the space train along the calculated trajectory, the expenditure of the working substance and the decrease of the general mass take place.

The variation process of the movement speed (V) and the mass (M) of the space train in the course of time is shown in Fig. 4.19.

The process begins from the motion of the train in the Earth's orbit around the Sun from the point 2 (Fig. 4.18). At this moment, Jupiter is also in its orbit on the position 0 and moves in 12 times more slowly than Earth. At the first part of the flight, the

Fig. 4.18 Planetary orbits
and the calculated flight
trajectory

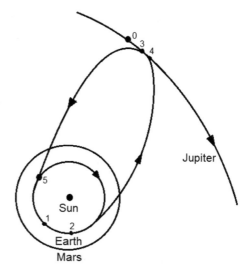

space train accelerates. The movement is accompanied by the speed increase and the decrease of the general mass. In 27 days, the flight speed of the space train reaches 400 km/s.

At the point 2, the train leaves the circular orbit and proceeds with the movement along the calculated trajectory, which has the form of parabola. The cruise rocket engines are switched off, and the working substance, which was there in the tank container, is already expended. In 47 days of the flight, the train approaches the planet Jupiter. By means of the thrust vector change on 180°, the braking of the space train, using the working substance, which is there in the tank container 4, is carried out. Decreasing the movement speed, the space train goes into the Jupiter orbit in 66 days of the flight. At this time, Jupiter, moving along its orbit, reaches the point 3, where already the space train is. The space train, maneuvering by means of electric rocket engines, crosses the orbits of Jupiter satellites Kallisto and Ganymede. Then, it makes one more maneuver and orbits Europe. After speed decrease up to 2 km/s, the space train carries out the orbital flight around Europe. Then, the space train divides.

The takeoff–landing capsule 4 comes off with the help of the docking unit 7. The astronauts, which are in the capsule, carry out the landing on the Europe surface. For this purpose, they switch on the chemical rocket engine 9 and move the capsule 4 into the braking mode. When the speed decreases lower than 2 km/s., the capsule loses weightlessness and is directed to the Europe surface.

With the help of the rocket engine 9, the soft landing in the chosen area on the Europe surface is carried out. (It shall be kept in mind that the gravity on the surface of Europe by 7,7 times less than on the surface of Earth). After the landing, the astronauts begin to carry out the research program, which is intended for 7 days.

For movement over the Europe surface, the astronauts use a delivered vehicle— "rocket sledge". At the same time, the space train, shown at Fig. 4.17, moves along

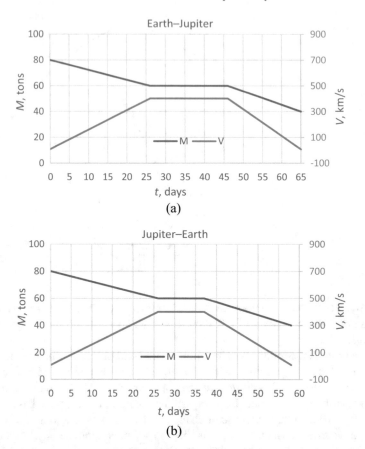

Fig. 4.19 Variation of movement speed (**a**) and the mass (**b**) of the space train

the Europe orbit with empty tank containers 3 and 4. The space refueler, which was delivered on the Europe surface, gets a starting signal.

All the elements, intended for working substance production on the Europe surface, sequentially join in the work. After completion of the working substance production cycle, the stand-alone refueler takes off the Europa surface and orbits it.

Being on the Europe orbit, the refueler approaches the space train and docks to it, as it shown in Fig. 4.16. The cryogenic pump is switched on, and the working substance is pumped over from tanks of the refueler into tank containers 4 and 3. After the completion of the refueling, the refueler 6 comes off from the train, leaves the Europe orbit, and carries out the landing on its surface.

The space train continues the movement along the Europe orbit already with completely filled tank containers 3 and 4.

After completion of intended research program of stay on Europe, the astronauts take the places in the crew cabin and switch on the chemical rocket engine. The takeoff–landing capsule begins to move straight up, and at achievement of the orbital velocity, it goes into the Europe orbit, along which the locomotive 1 (Fig. 4.15), docked to tank containers 2 and 3, moves.

After docking the capsule 4 to the tank container 3, the electric rocket engines 5 of the locomotive 1 are being switched on, and the last stage of the expedition—return into Earth's orbit—begins.

During the stay in the system of the planet Jupiter, the space train has moved along the Jupiter orbit from point 3 to point 4 (Fig. 4.18). The planet Earth, moving along its orbit, during the same period has passed from point 2 to point 1.

Now, the space train must move from point 4 into point 5 along the calculated trajectory, but in opposite direction, and as the Earth moves toward to the train, the return way will be by 340 million km shorter. And as it can be seen in Fig. 4.19, the flight duration from the Jupiter orbit into Earth's orbit decreases from 66 days down to 56 days. The flight between the Jupiter orbit and the Earth's orbit along the trajectory from point 4 to point 5 (Fig. 4.18) is carried out at the operation of electric rocket engine in acceleration mode, in the free flight mode, and in the brake mode at the rated power.

By going into the Earth's orbit, the space train (Fig. 4.15) is being divided. The takeoff–landing capsule 4 is separated from the tank container 3. The astronauts, which are in the cabin of the capsule, switch on the chemical rocket engine 9 and carry out a maneuver, resulting in the capsule going into the orbit of the International Space Station (ISS). The capsule 4 moors to the ISS and docks it. The astronauts leave the capsule 4 and pass into the ISS. After unloading the specimens, being delivered from the Europa surface, the speed flight, which has lasted 128 days, is over.

4.8   The Design of a Space Train to Fly to the Distant Planets of the Solar System

4.8.1   The Space Locomotive

The space locomotive 1 and the tank container with a working substance 2, which are connected with each other by the docking unit 3, are shown in Fig. 4.20.

Inside the space locomotive 1, a power installation, electric rocket engines, and a control system are located.

The space locomotive 1 has a cylindrical case with conic mouthpiece, which lengthwise is divided into two functional sections. In the tail-end section, the complex of the electric rocket engine is located. Four cruise electric rocket engines 5 are being moved out from the case of the locomotive 4 by means of the electric drive. Electric rocket engines 5 are located pairwise at an angle of 180° relative to each other. This allows to change the movement direction of the locomotive in space with the driving force change by electric current adjustment of cruise engine.

The working substance for the engines 5 is hydrogen, which is located in liquid state in a tank 6. In the engine compartment of the space locomotive, the flight control

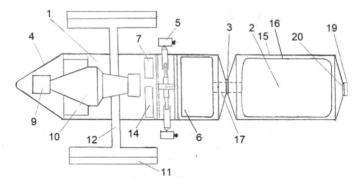

Fig. 4.20 Space locomotive design to fly distant planets of the Solar System

system 7 is located. The control system by means of the onboard computer and controllable power transistors regulates current of the cruise electric rocket engines 5.

The accumulator battery 14 is a part of the secured backup power supply system, which provides the independent control of the electrical installation in all operating modes.

In the fore part of the space locomotive 1, the onboard power installation, consisting of gas-phase nuclear reactor 9 and an energizer, is located. The energizer 10 is a MHD generator, mounted together with gas turbine generator along the locomotive axis. The cooling of the power installation is carried out by radiation into the space by means of a radiator 11, which is moved out the case of the space locomotive 4 by means of the holder 12.

The tank container with the working substance 2 has an external cylindrical-shaped cover 15. Inside the case, the vessel 16, in which the working substance (hydrogen in liquid state) is stored, is located. For heat gain decrease, the vessel 16 is covered by a layer of the screening vacuum isolation.

Liquid hydrogen comes from the tank container 2 through the pipeline 17 into the tank 6, which is located in the case of the locomotive 1.

After docking of the locomotive 1 with tank container 2 with the help of the docking assembly 3, both parts of the pipeline make a united hydraulic system. In the rear part of the tank container 2, the docking unit 19 is located, which makes it possible to connect the space tow with a takeoff–landing capsule. In order to provide the capsule rotation for the purpose to create the artificial gravitation in the astronauts' cabin, a bearing assembly 20 is mounted on the longitudinal axis of the space locomotive.

The calculated parameters and dimensions of the SERPS 5 rocket engine for the movement of rocket from Earth's orbit to the orbit of Jupiter are shown in Table 4.2.

Table 4.2 Parameters and dimensions of the SERPS 5 rocket engine

The electric rocket propulsion system parameters	
Traction force	250 N
Power	2500 kW
Current	2250 A
Voltage	1150 V
Efficiency	94.5%
Specific impulse	8000 s
Working substance consumption	3 g/s
Exhaust velocity of working substance	10 km/s
Magnetic induction	1.7 T
Anode diameter	170 mm
Cathode diameter	40 mm
Anode length	250 mm
Cathode length	100 mm
External cylinder diameter	900 mm
External cylinder length	600 mm

4.8.2 Onboard Power Installation

After consideration of all the possible ways of electric power production for power supply of the space electric motor in the project as the main source, the stand-alone nuclear reactor is assigned. The system of energy transformation with the gas heat carrier, which works according to the Brighton cycle, is developed. The system consists of two energy transforming blocks connected in series. The first block is AC MHD generator being located on the gas outlet from the reactor. After the MHD generator, the discharge gas arrives in the second block consisting of the turbine, the generator, and the compressor which are mounted on the same rotating shaft.

The main elements of the power system are shown in Fig. 4.21.

In the reactor 1, the heat, which is being transferred to the working substance (helium with additive of 1% of cesium) circulating in a closed gas contour, is being produced. The working substance heated to temperature of 2200 K passes through the MHD generator, where the mechanical energy of the ionized gas stream is transformed to the electric energy.

After the MHD generator, the working substance at temperature of 1600 K is being supplied on the input of the turbine 3, where gas thermal energy is transformed into the mechanical. The residues of the heat are discharged through the refrigerator radiator 6 in the open space at temperature of 1300 K. The cooled working substance is being compressed in the compressor 5 and then comes again at the input of the reactor 1.

The research conducted on a mathematical model has shown that the application of the developed thermal scheme will allow to increase the efficiency of the power

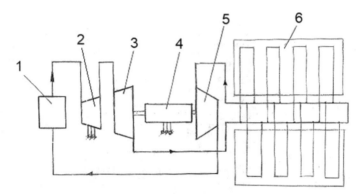

Fig. 4.21 Main components of the power system

installation up to 60%. Besides, the offered thermal scheme allows to increase the radiator temperature up to 1300 K and to decrease substantially its mass.

4.8.3   The Nuclear Reactor

In this project, the design of a nuclear reactor, which has been offered and developed in Brookhaven national laboratory of the USA [6], is built. The constructive scheme of the reactor is shown in Fig. 4.22.

It is a compact high-temperature reactor with fuel elements using the granular crushed particles of the spherical form having diameter 0,5 mm. The similar fuel elements are being produced out of uranium carbide and are used at the nuclear power plants of industrial power supply systems. The fuel elements 1 are being put

Fig. 4.22 Design of the nuclear reactor

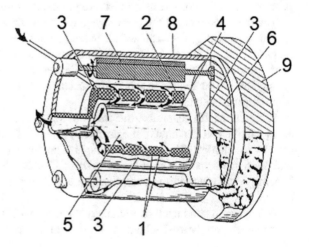

between two cylindrical drums 2, which are manufactured out of porous material on the basis of high-temperature ceramics.

The gaseous helium, which is used as the working substance, passes through the wall of an external drum 3, then through the filling 1, and the wall of an internal drum 4. The layer thickness of the granular nuclear fuel is 5 cm. Thus, a very large heat exchange surface is provided: 100 cm^2 on 1 cm^3 of the filling.

For improvement of neutron physical characteristics of the reactor in its central zone, a graphite column 5, which acts as a moderator reflector, is located. Thanks to intensive heat transfer from nuclear fuel to the heat transfer agent, the temperature difference between them is at the level of 12 K.

The reactor core is surrounded with a butt reflector 6 made of graphite in which cylindrical control drums 7 are located. On the outside, the nuclear reactor has the cylindrical load-bearing case 8 and the butt screen 9, which provides the protection against neutron flux. In Fig. 4.22, the movement of the heat carrier in the reactor is shown. The reactor critical mass is 30 kg of uranium. Diameter of the heat emission zone is 50 cm. The gas operating temperature at the input is 1300 K, at the output of 2200 K. The pressure of the heat carrier is 10 MPa, and the heating capacity of the reactor is 21 MW.

The control of the reactor is being carried out by means of drums 7. On one section of the drum the neutron absorber (boron carbide) and on the other section the moderator (graphite) are located. If at the drum rotation the section occupied with the absorber comes nearer to the fuel, the effective neutron multiplication factor is being decreased, and when the section occupied by the absorber moves away, the effective neutron multiplication factor is being increased. In stationary mode, the multiplication factor is one. In the scheme with external slowdown with the help of an absorber, the lifetime of neutrons is great, which allows you to work with slow change of reactive power at great reactivity. At the same time, the output at full capacity is carried out in a few seconds.

The nuclear reactor design, which is built into the project, completely satisfies the technical requirements [1] to power installations with the nuclear reactors being taken in space.

4.8.4   The MHD AC Generator

The MHD generator is installed aboard for transformation of the energy of the moving plasma stream into electric energy.

The analysis carried out has shown that all the existing MHD generators cannot provide the operation over the sufficient long period of time, i.e., to create the necessary resource. The reason is that the MHD generators of the traditional design have the essential shortcoming: the presence of metal electrodes. When the electric current is flowing between electrodes, the cathode is bombarded by ions of high energy, which leads to its corrosion. In addition, the design of the MHD generators is based on application of superconductors capable to create only a constant magnetic field. At

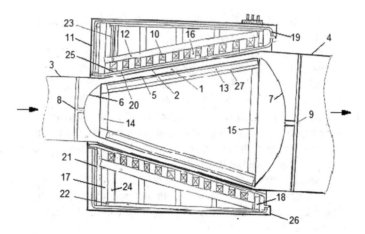

Fig. 4.23 Design of the MHD generator

present, the superconductors working on the alternating current are developed, and now, it is possible to create the alternating magnetic field at the minimum energy losses.

The MHD generator is proposed, which design excludes the application of electrodes that allows to provide electric power supply at long space flight. In the MHD generator, which design is shown in Fig. 4.23, the energy transformation is carried out by braking of a plasma stream in the traveling magnetic field.

From the nuclear reactor, the working substance in plasma state goes to the conic ring channel 1 on which external surface the three-phase AC winding 10 is placed. The AC winding 10 generates the magnetic field, which travels along the conic ring channel 1. The movement speed of the traveling magnetic field is made lower as compared with the movement speed of the working substance in the channel. Therefore, the traveling magnetic field induces in the electrically conducting working substance an EMF as well the eddy currents just as it takes place in the rotor of the linear asynchronous electric motor. At interaction of the eddy currents with the traveling magnetic field, the force comes into existence, which slows down the working substance flow and in the AC three-phase winding the EMF and the active current being delivered to the onboard power installation.

The speed of the traveling magnetic field is defined by frequency of the current in three-phase winding 10. In this connection, the electric scheme provides for the connection of the three-phase winding through the frequency converter.

The working chamber of the MHD generator 1 is formed by means of an external cone 2 and an internal cone 5 which are made of thermostatic ceramics. The end nozzles 6 and 7 provide the structural ruggedness of the internal cone 5. By means of holders of 8 and 9, the nozzles are fastened in the input tube 3 and in the output tube 4.

Table 4.3 MHD generator parameters

The MHD generator parameters	
The input diameter of the cone	0.6 m
The output diameter of the cone	0.9 m
The length of the working channel	1.1 m
The height of the working channel	0.06 m
The pole pitch of the winding	0.24 m
The speed of the working substance at the input	300 m/s
The speed of the working substance at the output	100 m/s
Active power	6000 kW
Voltage	3400 V
Current	2200 A
Current frequency	200 Hz
Magnetic induction	1.8 T
Power factor	0.45
Efficiency	0.82

The AC three-phase winding is being formed by connection of separate coils of the cylindrical form, each of which has different diameter. Installation of the bobbin winding 10 is carried out by means of cylindrical gaskets 16.

The cryostat 11, in which the AC superconducting winding is placed, has an outer casing covered with screening vacuum isolation, which provides minimum heat input into the area of the winding placement.

Liquid hydrogen, which is used as a coolant, is filled in the end chamber of the cryostat 21 and in the axial cylindrical chamber 22 which are used as the cryostat shields from external heat inputs. In the chamber 25, the liquid hydrogen cools the coils surfaces of the three-phase winding 10. The gaseous hydrogen being generated at boiling comes in the gas gathering chamber 26. For MHD generator calculation for the purpose of definition of its main parameters, the mathematical model in which the method of consecutive partitioning was used has been made. The calculation results are given in Table 4.3.

4.8.5   The Gas Turbine and the Cryoturbogenerator

The development of gas turbine, which is presented in the chapter, is based on the achievements of the company General Electric in the field of gas turbine manufacturing.

The main tendency is the increase of the initial gas temperature, which allows to get the maximal turbine efficiency. The achievement of the maximum temperature of the gas entering into the turbine (which level makes now 1600 K) is provided with

Table 4.4 Parameters of the cryoturbogenerator

Parameters	
Power	5000 kW
Voltage	3400 V
Current	1700 A
Power factor	0.5
Current frequency	200 Hz
Rotary speed	12,000 rpm
Magnetic induction	2 T
Efficiency	0.98

application of heat-resistant materials for the nozzle and operating turbine blades as well for turbine disks with simultaneous effective forced cooling of these details.

The turbogenerator was supposed to be used in the project of expedition to Mars—"Aelita", which was meant for power supply of electric rocket engines from a nuclear energy source.

The calculated values of the cryoturbogenerator parameters are given in Table 4.4.

4.8.6   The System of Electric Propulsion of the Space Train

The propulsion system, which was developed in the frames of the project, belongs to the class of electrical systems "generator—engine with stand-alone onboard nuclear energy source". In this case, the joint operation of the synchronous turbogenerator with MHD AC generator onto MHD DC engine was considered for the first time.

The description of the onboard power installation is given above.

The electric circuit of power installation, located in the locomotive, is shown in Fig. 4.24.

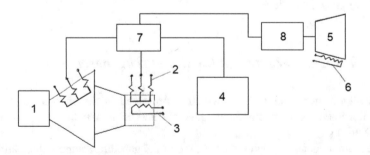

Fig. 4.24 Electric circuit of power installation

The special feature of the developed scheme consists in usage of the phenomenon of superconductivity for creation of magnetic fields in the main components of the installation.

The liquid hydrogen, which is located onboard of the locomotive, is used for cooling of the superconducting components and, simultaneously, as the working substance for the cruise electric rocket engines, which are installed onboard. The AC winding of the MHD generator 1 is connected to the AC winding of the turbogenerator 2 by means of the frequency converter 7. The excitation of the turbogenerator is carried out by means of the superconducting winding 3. The electric power consumer is the electric rocket engine 5, which has an external superconducting excitation coil 6.

For the power installation start-up, the storage battery 4 is placed onboard of the locomotive. The frequency converter 7 is performed on the basis of the controlled power transistors. The control system of the converter ensures the functioning of power installation in all the operating modes. For power installation start-up, the constant voltage from the storage battery 4 is fed to the input of the frequency converter 7, which is connected to the armature winding of the turbogenerator 2.

The frequency converter produces a current, which frequency gradually increases. After switching on of the excitation coil of the turbogenerator 3, it is accelerated up to nominal speed, working as a synchronous electric motor.

Simultaneously, the compressor, which is pumping over the gaseous helium in the closed circulation loop of the power installation, begins to operate. Then, the start of the gas-phase nuclear reactor (by means of control drums) takes place.

When the pressure and gas temperature are being increased, the turbine, which provides the energy for the compressor operation, starts to work. The accumulator battery becomes disconnected from the convertor 7. For start-up of the MHD generator, the frequency converter 7 is connected to the winding 1. Then, the current in the excitation coil of the turbogenerator increases, and the turbogenerator begins to operate as a synchronous compensator.

Electric current from the winding of the turbogenerator 2 through the frequency converter 7 comes in the winding 1 of the MHD generator. In working channel of the MHD generator, the traveling magnetic field is being generated, and it produces an active power.

The electric rocket engine 5 is connected to the power installation through the controlled rectifier 8, which carries out the control of the cruise engines 5. By changing current in the electric rocket engine, the traction force of each of four cruise engines is being changed, providing necessary flight direction of the space locomotive. As an explanation, in Table 4.5, the main data of the components, which are shown in Fig. 4.24, are given.

The calculations showed that flights to the planets of the Solar System should be carried out by using two different electric propulsion systems:

– For flights to the planets Mars, Venus, and Mercury, the electric propulsion system is developed, in which the assembly module is a rocket moving with the help of electric rocket engines. The source of energy for the engines is a sliding solar

Table 4.5 Characteristics of electric propulsion system

Characteristics	
MHD generator	
Power	6000 kW
Voltage	3400 V
Current	2200 A
cos φ	0,45
Turbogenerator	
Power	5000 kW
Voltage	3400 V
Current	1700 A
cos φ	0,5
MHD engine	
Power	2500 kW
Voltage	1150 V
Current	2250 A
Tractive force	250 N
Storage battery	
Voltage	12 V
Capacity	4000 Ah

panel based on gallium arsenide. The working substance for the electric rocket engines is nitrogen, which is stored in liquid state in a cryogenic tank located along the rocket axis. When entering the Earth's orbit, a takeoff–landing capsule with a cockpit for the crew connects to the rocket.

– For flights to the planets Jupiter, Saturn, Uranus, and Neptune, an electrical propulsion system is developed, in which the rocket consists of several serially connected modules. The first module is a space locomotive, inside which onboard power plant, electric rocket engines, and control system are located. The electric power unit, which feeds the SERPS-type electric rocket engines, consists of a gas-phase nuclear reactor, a magnetohydrodynamic generator, and a turbogenerator. The second module of the rocket is a tank container with the working substance hydrogen, which is stored in liquid state. The third module is a takeoff–landing capsule with a crew cabin.

4.9   Conclusion

After the project development of the space train for the expedition to Jupiter, the projects of space trains for expeditions to the planets Saturn and Uranus have been carried out. The whole cycle of calculation and design work, determining the possibility of space exploration in the twenty-first century ends with the expedition toward Neptune.

While summarizing, one can tell that an engineering decision of the global task of interplanetary space mastering has been found, which makes it possible to realize the expeditions toward all the planets of the Solar System.

The completed works show that expeditions should be carried out in stages, starting with Mars and Jupiter. For further expeditions to Saturn, Uranus, and Neptune, a method has proposed and developed, which is that to ensure a minimum time flight from Earth's orbit to the orbit of the gas planet used a "space refueler" of working substance, which is pre-delivered to the surface of the satellite of the planet, covered with a layer of water ice.

The works show that the most appropriate technical solution of the expedition delivery means is a space train driven by a locomotive equipped by an electric rocket engine. The electric rocket engine of magnetoplasma type with an external superconducting winding, which creates a transverse magnetic field, is connected to the onboard power plant consisting of a nuclear reactor, an MHD AC generator, and a cryoturbogenerator. The working substance of the electric motor, liquid hydrogen, is placed in tank containers connected to the locomotive by docking units. The crew of expeditions is located in the cabin of the takeoff–landing capsule, which is connected to the tank containers. The results of the calculation and design studies are presented in Table 4.6.

Table 4.6 captures the situation that will be formed, if the expedition to the planet Neptune is successfully carried out.

Technical means have been created that allow surmounting the maximum distance from Earth up to 29 AU. The space refuelers installed on Europe, Titan, Oberon, and Triton will allow to carry out repeated expeditions and regular flights to all planets of the Solar System without any prior flights. Thus, at the very beginning of the twenty-first century, a system of interplanetary communications was actually designed, which will be improved over time. It should be noted once again that the space refuelers are able to fly on command from Earth around the satellites on the surfaces of which they are installed. Therefore, the refuelers can serve as permanent research laboratories.

Table 4.6 shows that the maximum duration of flight from the Earth's orbit to the boundary of the Solar System will not exceed 90 days. The maximum speed of the space train when flying to distant planets will reach 900 km/s, and the maximum power of the onboard power plant will reach 40 MW. Today, this limit reflects the real level of space technology.

Table 4.6 Characteristics of expeditions to the planets of the solar system

Planet	Mercury	Venus	Mars	Jupiter	Saturn	Uranus	Neptune
Distance from Earth (A.U.)	0.6	0.28	0.64	4.4	9	19	29
Place of refueling	Earth	Earth	Mars	Europa	Titan	Oberon	Triton
Carrier rocket	Delta heavy	Arian 5	Delta heavy	Arian 5	Delta havy	Delta heavy	Falcon heavy
Flight duration (days)	36	21	40	65	68	88	91
Number of tanks with working substance	1	1	2	2	2	3	3
Tractive force of locomotive (kN)	0.64	0.5	0.5	1.0	2.0	2.0	4.0
Power of onboard energy unit (MW)	6	3	3	10	20	20	40
Max. speed (km/s)	130	120	100	400	700	900	900
Energy unit	Solar battery	Solar battery	Solar battery	Nuclear reactor	Nuclear reactor	Nuclear reactor	Nuclear reactor

References

1. D. Buden, J.A. Sullivan, *Nuclear space power systems for orbit raising and maneuvering.* *American Institute of Aeronautics and Astronautics* (New York, NY, 1984)
2. E. Choueiri, The efficient future of deep-space travel-electric rockets; Das Zeitalter der Elektrischen Raketen Das Zeitalter der elektrischen Raketen. Spektrum der Wissenschaft **1**, 32–39 (2010). ((In German))
3. R.A. Coombe, *Magnetohydrodynamic Generation of Electrical Power* (Chapman and Hall, London, 1964)
4. S.B. Gabriel Energy storage systems for MPD thrusters, AIAA, No 81–0142, (California Institute of Technology, Jet Propulsion Laboratory, Pasadena, California, 1981)
5. V.P. Konovalov, E.E. Son, Electric conductivity of molecular hydrogen plasma with alkali metal additive. High Temp **49**, 138–140 (2011)
6. B. Powell, Particle bed reactors and related concept. In *Proc. of Symposium of Advanced Compact Reactors Systems* (Washington DC, USA, November 1982)
7. A.E. Roy, *Orbital Motion* (CRC Press, Boca Raton, Florida, 2004)
8. A. Rubinraut, The study of the Electric Rocket Engine for the Future. Advances in Aerospace Science and Technology **2**(1), 1–16 (2017)
9. I.D. Urusov, *MHD Generators* (Science, Moscow (in Russian), 1966)

Printed in the United States
by Baker & Taylor Publisher Services